MYANMAR
THAILAND
Bangkok
man Sea
huket
Penang
MALAYSIA
Kuala Lumpur
ledan
SINGAPORE
Sumatra
INDONESIA
Jakarta

AF685432

ACEH

A New Dawn

The **Institute of Southeast Asian Studies (ISEAS)** was established as an autonomous organization in 1968. It is a regional centre dedicated to the study of socio-political, security and economic trends and developments in Southeast Asia and its wider geostrategic and economic environment.

The Institute's research programmes are the Regional Economic Studies (RES, including ASEAN and APEC), Regional Strategic and Political Studies (RSPS), and Regional Social and Cultural Studies (RSCS).

ISEAS Publishing, an established academic press, has issued almost 2,000 books and journals. It is the largest scholarly publisher of research about Southeast Asia from within the region. ISEAS Publishing works with many other academic and trade publishers and distributors to disseminate important research and analyses from and about Southeast Asia to the rest of the world.

ACEH
A New Dawn

EDITED BY

ARIS ANANTA & LEE POH ONN

ISEAS

INSTITUTE OF SOUTHEAST ASIAN STUDIES

Singapore

First published in Singapore in 2007 by ISEAS Publishing
Institute of Southeast Asian Studies
30 Heng Mui Keng Terrace
Pasir Panjang
Singapore 119614

E-mail: publish@iseas.edu.sg
Website: <http://bookshop.iseas.edu.sg>

ISEAS Library Cataloguing-in-Publication Data

Aceh: a new dawn / edited by Aris Ananta and Lee Poh Onn.
1. Aceh (Indonesia)—Politics and government.
2. Natural disasters—Indonesia—Aceh.
3. Aceh (Indonesia)—Population.
4. Ethnicity—Indonesia—Aceh.
5. Aceh (Indonesia)—Religion.
6. Women—Indonesia—Aceh.
7. Disaster relief—Singapore.
8. Disaster relief—Indonesia—Meulaboh (Aceh).
I. Ananta, Aris.
II. Lee, Poh Onn.
DS646.15 A1A173 2007

ISBN-13: 978-981-230-395-0 (hard cover — 13 digit)
ISBN-10: 981-230-395-2 (hard cover — 10 digit)

Endpaper of map is reproduced from *Reaching Out: Operation Flying Eagle, SAF Humanitarian Assistance after the Tsunami* (Singapore: SNP Editions for Ministry of Defence, 2005) with kind permission of the publisher.

Typeset by Superskill Graphics Pte Ltd, Singapore
Printed in Singapore by Utopia Press Pte Ltd

Contents

Foreword

The Asian tsunami of 2004 was one of the deadliest disasters in modern history. It caused unprecedented death and destruction, and suffering to the people of the affected nations. It was also a grim reminder of the havoc Mother Nature can unleash. In the midst of the tragedy, we also saw the world, with Singapore at the forefront, band together to offer emergency aid to the victims.

While much has been written about this disaster, the work that has continued since the devastation struck has not been given sufficient attention. Two years on, it is timely to review what has been done to bring the victims and their nations back on their feet and to recognize the efforts of all who responded to the call for assistance in the immediate aftermath of the disaster, and later in the recovery and reconstruction stages.

When the Director of ISEAS, Ambassador K. Kesavapany, approached me with the idea of producing a book to commemorate the second anniversary of this event, I did not hesitate and assured him of my full support. Besides Aceh as a whole, this book also focuses on Meulaboh, Indonesia, as this was the region most affected by the disaster. It gives an insight to the speedy, focused, and appropriate response by the Singapore Government and the efforts by Singaporeans and our NGOs in the reconstruction and rehabilitation process which is still on-going. While this book is targeted at Singapore's efforts in Meulaboh, Aceh, it is necessary to point out that we also reached out and provided significant assistance during the emergency through to the reconstruction stages in Sri Lanka and the Maldives. Many of Singapore's NGOs and VWOs were and still are in these two countries to bring humanitarian relief to those affected by the tsunami.

Dr Aris Ananta and Dr Lee Poh Onn, scholars at the Institute of Southeast Asian Studies, have covered a lot of ground in bringing this book together. I have been to Meulaboh and seen for myself the devastation immediately following the tsunami as well as the recovery efforts during my subsequent visits. The work of these two scholars reflects and confirms what I have seen of the work of all the Singaporeans who have made the special effort and sacrifice to provide assistance to the people of Meulaboh. This book has done justice in recording the sterling efforts of Singaporeans in lending a helping hand to their neighbours in times of need.

Lt Gen Winston Choo, Retd.
Chairman
Singapore Red Cross

Message from Pieter Feith

The Link between Reconstruction and Peace

Even though the first contacts between the parties in the Aceh conflict had already taken place when the December 2004 tsunami struck, the disaster brought new urgency with it: a political will to leave old grievances behind and join forces in the necessary reconstruction process and the creation of a common and sustainable future for the people of Aceh.

There is an intimate link between peace and development, which in the case of Aceh translates into the interdependence between post-tsunami reconstruction and the peace process. It has been stated before, but it is essential to reiterate that without security and political stability the continued rebuilding of the Acehnese society and economy would be extremely difficult — an observation reiterated in the preamble to the Memorandum of Understanding signed by the parties in Helsinki on 15 August 2005. Conversely, reintegration of former GAM (Free Aceh Movement) combatants is helped by the massive reconstruction effort in Aceh, strongly supported by the European Union. After the Aceh Monitoring Mission (AMM) leaves Aceh, the reintegration efforts for the victims of the conflict, as well as the reconstruction of houses and infrastructure ruined by the tsunami or by the conflict will have to be embedded in longer-term programmes benefiting all affected citizens of Aceh.

There is a lot at stake for the people of Aceh. Keeping the tsunami tragedy in mind, the EU is a long-term partner to the people of Aceh as they progress along the irreversible path to sustainable peace and development in Aceh.

Pieter Feith
Head
Aceh Monitoring Mission

Message from the Director

The study of Indonesia has remained one the most important focal points at the Institute of Southeast Asian Studies. In this context, the regional autonomy policies implemented in Indonesia since 2001 have generated a dynamism of their own, given that regions and districts now exercise a greater autonomy in the decision-making process. This has in turn initiated new approaches to studying the developments in Indonesia. Likewise at ISEAS our studies of Indonesia have taken a regional approach beginning with the Riau Archipelago and following up with Nanggroe Aceh Darussalam, both within the island of Sumatra.

This book is the first output of the study on Nanggroe Aceh Darussalam, edited by Aris Ananta and Lee Poh Onn. All of the authors here are either research staff (Aris Ananta and Lee Poh Onn) or visiting research fellows (Michael Vatikiotis and Braema Mathiaparanam) at ISEAS, with the exception of Bapak Kuntoro Mangkusubroto, Head of the BRR (Rehabilitation and Reconstruction Agency), who has graciously contributed the chapter on the role of the BRR in the recovery and reconstruction process in Aceh and Nias.

This book examines Aceh from a new beginning, since the occurrence of the tsunami in December 2004. A new Aceh, with a new dawn to build lasting peaceful and prosperous development in Nanggroe Aceh Darussalam.

Finally, I commend the two editors and all the authors for their seriousness in understanding the importance of the current issues to Aceh. I would also like to acknowledge the contribution of Bapak Kuntoro Mangkusubroto, who presented an ISEAS public lecture on "The Situation in Aceh and Nias: An Update" on 2 March 2006, and which has subsequently been revised and included as a chapter in this book. I also appreciate the hard work of Mrs Triena Ong and her staff in the Publications Unit for producing this attractive publication.

K. Kesavapany
Director
Institute of Southeast Asian Studies
Singapore

The Contributors

Aris Ananta is Senior Research Fellow, Institute of Southeast Asian Studies, Singapore

Lee Poh Onn is Fellow, Institute of Southeast Asian Studies, Singapore

Kuntoro Mangkusobroto is Director of Badan Rehabilitasi dan Rekonstruksi, BRR (Rehabilitation and Reconstruction Agency) NAD-Nias, Banda Aceh, Indonesia

Braema Mathiaparanam is Visiting Research Fellow, Institute of Southeast Asian Studies, Singapore

Michael Vatikiotis is Visiting Fellow, Institute of Southeast Asian Studies, Singapore

Introduction

Aris Ananta
Lee Poh Onn

Well-known for its beautiful beaches, West Aceh faces the Indian Ocean. It is in the province of Nanggroe Aceh Darussalam (NAD) located on the northern tip of the large island of Sumatra in the Indonesian archipelago. With a tropical climate and heavy rainfall, the district used to have vast areas of fertile land. The city of Meulaboh, its capital, was historically a busy transit point for business ships from many countries. However, during modern times, this district has also often suffered from natural disasters such as floods, typhoons, and landslides. Some of the major disasters occurred in 1962, 1978, 1992, and 1999.[1]

None, however, had reached the magnitude of the earthquake and tsunami which struck the western coast of NAD in the morning of 26 December 2004, totally changing the landscape of West Aceh, particularly the city of Meulaboh.[2] Initially, the earthquake struck 150 kilometres off the coast of Aceh. Forty-five minutes later, the tsunami hit Aceh destroying approximately 800 kilometres of its coastline. In Aceh, over 127,000 houses were totally damaged or partly destroyed. Over 500,000 people were made homeless within an instance.

Livelihoods were also severely disrupted. The disaster brought heavy ruin to important economic sectors, in

Facing page: Destroyed town. *Photo courtesy of Mercy Relief.*

Photo courtesy of Mercy Relief

Exhausted rescue worker.

particular fisheries, agriculture, and small enterprises. The fisheries sector was estimated to have suffered the most, with a loss of US$511 million, followed by agriculture (US$225 million), and small enterprises (US$218 million). Two-thirds of all boats were damaged, destroyed, or lost. Large areas of land and fish ponds were completely wiped out. Agricultural land also suffered serious seawater inundation, with the tsunami destroying over 28,000 hectares of plantations. Marketing and extension services to support fishing and agriculture were severely disrupted, with many still in the process of recovery. Many small and medium enterprises (SMEs) lost their assets.

In terms of infrastructure, six hospitals, 41 *puskesmas* (community health centres), 59 *pustu* (auxiliary community health centres), 44 *posyandu* (integrated health service posts), and 240 *polindes* (village delivery posts) were severely damaged or destroyed. Major water treatment installations, the piping network and sanitation facilities were completely damaged. Along the west coast of NAD, extensive portions of the road network and bridges were destroyed. Most seaports, airports, ferry terminals, and inter-island boat stations were destroyed. More than 2,000 schools were damaged including textbooks and learning materials. About 2,500 teachers and 40,000 students lost their lives.

In Meulaboh, the coastal line was completely destroyed. All roads and bridges, including those connecting Meulaboh to the city of Medan, the capital of the neighbouring province of North Sumatra, and city of Banda Aceh, the capital of the province of NAD, were severely damaged or made impassable. There was neither electricity nor telephone communication (including cellular telephone). There was total darkness at night. Ruined structures and crops were everywhere. Weeping and crying filled the air. About 50 per cent of the population of Meulaboh of around 176,586 lost their homes. Meulaboh became an isolated city.

The destroyed and damaged infrastructure and communication system including transportation hampered the relief efforts and subsequent recovery and reconstruction process. Initially, the only way to supply help, including medical facilities and foods, was by air, undertaken by Chinook helicopters provided by the Singapore Government. The Singapore Armed Forces were the first group to come to help the survivors of Meulaboh.

Together with NGOs, such as Mercy Relief, the Singapore Red Cross, and the Singapore International Foundation, the Singapore Government provided significant assistance in the early relief and the following recovery and reconstruction, particularly in Banda Aceh and Meulaboh. The Singapore Government, Singapore Red Cross and Temasek Holdings Private Limited contributed to the construction of the pier at the Meulaboh Harbour, which plays a very strategic role in transporting building materials and supplies for the recovery and reconstruction process.

Currently, the Meulaboh Pier is also the only operational pier along the west coast of NAD. With the operation of the pier, the recovery and reconstruction efforts do not have to depend on the road infrastructure, which was mostly destroyed. The pier is especially important because it paves the way for aid to pour into Meulaboh as well as other areas in NAD that suffered from the onslaught of the tsunami.

Meulaboh today is very different from the period immediately following the disaster, thanks to the help and cooperation from the Indonesian and Singapore Governments, their citizens, and NGOs, as well as other international communities. However, although efforts were concentrated in Meulaboh,

Singapore teams were also very active in Banda Aceh and other parts of NAD. The Singapore agencies involved the people of Singapore at many levels. Singaporeans gave donations and collected food, medicines, blankets, and clothing especially during the initial period after the tsunami. The Government of Singapore supplied logistics and support for the relief effort, and also offered its air and naval bases. Prime Minister Lee Hsien Loong initiated the special summit of ASEAN leaders days after the occurrence of the tsunami. During the summit, the leaders requested the United Nations to establish a special fund to help post-disaster relief and reconstruction efforts.

Realizing that all efforts in the relief, recovery, reconstruction, and long-term development are oriented towards the needs of the community and implemented in a region which had violent conflicts for almost 30 years, it is hoped that this book will contribute to a better understanding of the overall

Photo courtesy of Mercy Relief

Books water-logged and damaged by the tsunami.

sustainable development process in NAD. The book examines issues related to not only natural disasters but also violent conflict — both of which have afflicted Aceh significantly in recent years. Chapter 1 begins with a discussion on the fast changes happening in NAD after the earthquakes and tsunami. Michael Vatikiotis, Visiting Fellow at ISEAS, discusses a promising new beginning for NAD.

The community-based orientation of the recovery and reconstruction process must first involve a better understanding of the characteristics of the population, as both the important subjects and objects of the long-term development in NAD. In Chapter 2, Aris Ananta, Senior Research Fellow at ISEAS, examines some population characteristics in both the natural disaster and conflict areas. The discussion in this chapter benefits directly from the recently available post-tsunami population census. It focuses on the provincial and district levels, especially the geography of the impact of the natural disasters and conflicts.

Chapters 3, 4 and 5 narrow the discussion to issues specifically related to the natural disaster. The progress and challenges in the rehabilitation and reconstruction of NAD are presented by Kuntoro Mangkusubroto, Director of the Rehabilitation and Reconstruction Agency (BRR). This chapter is based on his presentation on "The Situation in Aceh and Nias: An Update", at the public lecture jointly organized by the Institute of Southeast Asian Studies (ISEAS) and the Lee Kuan Yew Exchange Fellowship, on 2 March 2006 in Singapore. In Chapter 4, Braema Mathiaparanam, Visiting Fellow at ISEAS, concentrates on a special and important issue of development involving the impact of the natural disaster on women. The contribution of the Singapore Government, its people and NGOs in the process of relief, recovery and reconstruction is discussed in Chapter 5, by Lee Poh Onn, Fellow at ISEAS. The book ends with concluding remarks, containing policy recommendations for lasting peace and prosperity in NAD.[3]

Notes

1. Tim Litbang, *Profil Daerah Kabupaten dan Kota. Kompas*, vol. 2 (Jakarta: Penerbit Buku Kompas, 2003), pp. 17–18.
2. Data on the extent of the earthquakes and tsunami draws heavily from BRR NAD-NIAS, *Aceh and Nias One Year After the Tsunami: The Recovery Effort and Way Forward* (Banda Aceh, Indonesia: Badan Rehabilitasi dan Rekonstruksi NAD-Nias, 2005).
3. We wish to thank Col Christopher Chua (Retd), Senior Operations Manager of the Singapore Red Cross Society, for comments on an earlier draft of this monograph. The normal caveats apply; all remaining errors are solely attributable to the editors alone.

1

Civil War, Conflicts and Natural Disasters

Michael Vatikiotis

The tsunami that hit Aceh on 26 December 2004 was a natural disaster of historic proportions. More than 170,000 people lost their lives. Great swathes of coastline were submerged and mauled by the earthquake and subsequent giant black waves that hit without warning that eerily bright and clear morning. Half a million people were left homeless; 600 villages were destroyed. The city of Banda Aceh resembled a nuclear bombsite, its buildings flattened. Only the great city mosque was left standing in the area washed over by the waves. Driving through the smaller town of Meulaboh just a few weeks later gave one the curious sense of being on the film set of a disaster movie.

Yet the tsunami will be remembered as a historical turning point as well. For in the wake of the devastating black waves, there followed a peace agreement with the rebel movement that had waged war with Jakarta for thirty years. It took less than seven months for the government in Jakarta and the Free Aceh Movement based in Sweden to reach a

Facing page: Wreckage in Banda Aceh. *Photo courtesy of Mercy Relief.*

deal and sign a Memorandum of Understanding in Helsinki in August 2005. In the subsequent months, as sceptics predicted that Jakarta would have a hard time selling the agreement at home, or that the armed rebels would not surrender their weapons, it became clear that the tsunami had exhausted both sides and there was a willingness to work towards peace. From the ground, amid the wreckage of homes, schools, and hospitals, and the memories of so many vanished lives, it was very hard to imagine anyone arguing for more war.

One thing was certain after the tsunami that killed so many tens of thousands of Acehnese: there's nothing more pressing for ordinary people than the need to rebuild their shattered lives. It was never easy on the ground to elicit views about autonomy or independence; the security forces were everywhere, as was the fear of reprisal or extortion from the Free Aceh Movement. Yet after decades of civil war, a devastating natural disaster, and now with the promise of billions of dollars in aid for reconstruction, all the Acehnese really seemed to want was a normal life. The major disaster had changed the mindsets of local and national leaders alike to end the conflict.

* * *

The roots of conflict in Aceh stretch back into the Indonesian republic's early history. The modern claim to Acehnese independence is based on the fact that the old Islamic sultanate with its trading links to Europe and the Middle East was never formally incorporated into the Dutch colonial empire. Yet the Acehnese agreed to join the fledgling republic in 1949 in return for special treatment. Jakarta gave Aceh "special territory" status, but then proceeded to carve up the area's abundant natural resources between foreign and Jakarta-based interests. After an early rebellion was put down in the 1950s, a new separatist movement emerged in the mid-1970s.

Initially, the "Gerakan Aceh Merdeka" (GAM), or Free Aceh Movement, could only draw on limited popular support. But as the Indonesian military reaction to the movement grew harsher, so too grew the alienation of ordinary Acehnese. The Indonesian military's common practice of "sweeping" for rebels meant that ordinary Acehnese villagers were subjected to rough interrogation and random acts of violence. Although pushed underground, the GAM could draw on popular support in remote areas and by the late 1990s even in some urban areas. The Indonesian military invested heavily in special operations, but never managed to defeat the rebels. More than 15,000 people, mostly civilians, died in the protracted rebellion. During the brief cessation of hostilities in 2002–03, organized and implemented by the Geneva-based Centre for Humanitarian Dialogue, it is widely recognized that the level of conflict was reduced and human rights abuses on both sides curtailed. But there was precious little goodwill on either side. The GAM was unwilling to give up its claim to independence, and the Government of Indonesia was under pressure from security quarters to crush rather than deal with the rebels. Even under the enlightened government of President Abdurrahman Wahid in 2001–02, there was almost no chance that a negotiated settlement would survive in Jakarta's factionalized and nationalistic political environment.

Then the tsunami hit Aceh and everything changed. Perhaps one of the

After the tsunami, Aceh welcomed outsiders, including members from the international community.

Photo courtesy of Mercy Relief

most important turning points was when newly installed President Susilo Bambang Yudhoyono decided to allow, with almost no hesitation, thousands of foreign military and civilian forces to bring relief to Aceh. It was a remarkable sight to behold. What had been, just a few weeks before, a closed province under martial law, was suddenly transformed into a massive international relief operation. The airport at Banda Aceh resembled a vast international trade fair, with personnel and equipment from countries as far away as China, Spain, and Russia. Military helicopters from Singapore, Malaysia, the United Kingdom and the United States circled overhead, almost landing on top of one another. Singapore's own Chinook heavy-lift helicopters proved invaluable — so much so that the Indonesian Armed Forces has placed a priority on acquiring some of these giant twin-rotor machines.

The massive relief operation made history, too. For Indonesia is a proud nation that has not always accepted offers of help from its neighbours with such open arms. When in 1998 Malaysian fire-fighters sailed to the coast of Sumatra to help put out bush fires that were generating a suffocating haze affecting the Malaysian peninsula, senior Indonesian officials almost sent them back. In Aceh, the scale of the disaster ensured that things were different. There was also a genuine outpouring of charity and volunteer labour in countries like Singapore, where just five years earlier many were reluctant to travel to Jakarta as tourists because of fears about ethnic and religious violence. It took a tsunami to break down residual ethnic and religious barriers to aid and cooperation flowing freely across the Malacca Straits.

In these circumstances it was hard to imagine anything other than a return to the negotiating table. One stroke of luck, and an equally important precursor to peace, was the fact that Indonesia's new President had also been a supporter of the earlier attempt to seek a negotiated solution under the Cessation of Hostilities Agreement (COHA). His vice-president, Yusuf Kalla, had earned his own spurs as a peacemaker in Maluku, where he had helped broker an end to a vicious outbreak of religious violence in 1999. It was Kalla's own business network that helped lead the government to the Conflict Management Institute, a small conflict resolution outfit based in Helsinki that took up the baton of mediation from Henry Dunant Centre's (HDC) earlier groundbreaking efforts.

Having said all this, the tsunami effect will not guarantee a lasting peace. Critical to the success of the peace agreement was the show of goodwill from Jakarta in terms of a political commitment not only to implementing special autonomy arrangements under the peace agreement, but also to rebuilding the province.

* * *

Photo courtesy of Mercy Relief

Supplies pouring into Aceh.

When Kuntoro Mangkusubroto was appointed to head the Aceh Rehabilitation and Reconstruction Agency tasked with rebuilding Aceh, there were those in Jakarta who said that if this conspicuously clean former Minister for Mines and Energy could handle Aceh, then he could surely fix the rest of the country. It was a measure of how much faith was placed in one man to manage Aceh's recovery — and also of how much of Indonesia's progress was to be seen through the lens of Aceh. More than one year on and it is hard to find anyone who would disagree that Kuntoro has risen to the challenge. To be sure, almost 70,000 people still remain in tents, only one third of the housing needed has been built. But no one would quibble with how long it takes to rebuild shattered infrastructure properly; where there are fears is how much of the money will be spent wisely and honestly. And here there has been almost no serious accusation of embezzlement or corruption, which was everyone's big concern.

Managing US$7 billion in aid puts Kuntoro and his team on the same level as a national technocrat dealing with national-level investment. Only in Aceh, the stakes are even higher. For if Jakarta fails to rebuild enough homes and schools, fails to convince the Acehnese of its goodwill and honest intentions, there is no doubt that renewed rebellion remains an option. Here lies the longer-term challenge in Aceh. Kuntoro, the popular and incorruptible technocrat, will not be around forever. At some point a new locally elected government will assume the reigns of reconstruction and development. Elections under the new autonomy arrangements will have taken place in December 2006, the peace agreement monitors from ASEAN and the EU will leave. Then what?

The biggest fear for the future is that Indonesia's central government will not really let go. Aceh's special autonomy arrangement, in its new form, is a test bed for substantive decentralization, not just of the functions of government, but also of the selfish business interests that have prevented profits from local natural resources returning to local people. New national laws on decentralization in place since 1999 now foster a much greater degree of local autonomy across Indonesia, and in many ways the speed of decentralization has thrown up the counter challenge of how to ensure that Indonesia's far flung parts have the funds to develop and support the welfare programmes once nurtured by central authority.

The more that this creeping form of federalism takes hold, the more intensely a political counter-reaction will develop in Jakarta. Nationalist elements in the Indonesian parliament are convinced that Indonesian negotiators signed away too much autonomy to the Acehnese in Helsinki. The Memorandum of Understanding signed in August 2005 omits to describe Aceh as a "province" of Indonesia; it calls for a head of government rather than a governor. Much of this language is contentious and may well be changed by the time the new law on special autonomy is passed. The Free Aceh Movement is demanding that the government abides by the agreement, but its leaders are exhausted and have returned to Indonesia; they will be satisfied with a form of autonomy that allows them to contest local politics and lead Aceh from the ground. Despite the wrangling in Jakarta, there is a high probability that the agreement will stick.

The legacy of the tsunami will ensure that Aceh remains a special case, demanding special attention for years to come. But longer term, a decade from now, the real test of all the attention that has been lavished on Aceh since December 2004 will be whether or not this potentially prosperous northern region of Sumatra, with abundant natural resources, proximity to the rest of Southeast Asia and situated on strategic sea-lanes to and from East and West, will emerge as a significant zone of prosperity. Only time will tell.

Photo courtesy of Mercy Relief

School children receiving donations of educational materials.

2

The Population and Conflicts

Aris Ananta[1]

Natural Disasters in the West, Conflicts in the East

The province of NAD (Nanggroe Aceh Darussalam) is one of the most resource-rich provinces in Indonesia. In 2003, natural resources (oil, gas, and their products) contributed about 30.0 per cent of the province's gross regional domestic product.[2] The tsunami obviously affected economic output of this province, though the negative impact has varied from one district to another. The largest estimated economic loss was in Aceh Jaya (86.4 per cent) and Simeulue (56.0 per cent). Yet the districts of North Aceh and Aceh Tamiang suffered very little: their gross regional products were estimated to decline by around 0.5 per cent only.[3]

The contribution of the province's gross regional domestic product to that of Indonesia as whole was small — only 2.26 per cent in 2003. Without oil and gas, the contribution was even smaller, 1.43 per cent.[4] Therefore, the negative impact of the tsunami on the total Indonesian economic output was very small.

As in other resource-rich provinces, there has been a long dispute with Jakarta (central government) over the

Facing page: Queueing for food at an aid centre. *Photo courtesy of Mercy Relief.*

actual distribution of economic income between Jakarta and the province. Furthermore, GAM (Gerakan Aceh Merdeka or Free Aceh Movement) had been fighting for the independence of the province from the Republic of Indonesia since the 1970s. The fighting was based more on ethno-nationalism, rather than an establishment of an Islamic state in Aceh. Violent conflicts had resulted in heavy losses in terms of both human casualties and economic damage.

Interestingly, the tsunami disaster provided an opportunity to end the conflict and to create peace in NAD. After the tsunami, conflicts became almost non-existent. The Government of Indonesia and GAM eventually signed a peace agreement, or "Helsinki Memorandum of Understanding", on 15 August 2005, followed by the issuance of the Law on Governing Aceh on 11 July 2006.

The province is located in the north-western tip of Sumatra. It is surrounded by the sea: the Indian Ocean on the west and southwest, and the Malacca Strait on the north and northeast. It borders with the province of North Sumatra on the southeast (see Figure 2.1). The western coast of the province was the hardest hit during the tsunami, especially those districts further to the north, such as the city of Banda Aceh, very near to the epicentre of the earthquake/tsunami. Meulaboh, the capital of the regency of West Aceh, approximately 150 km from the epicentre of the earthquake, was also terribly devastated, more so than areas further south.

Historically, most of the districts associated with GAM were on the east coast particularly in the regencies of Pidie, Bireuen, North Aceh, East Aceh, Aceh Tamiang and the cities of Lhokseumawe and Langsa. All of them were relatively safe from the natural disaster. The regency of Aceh Besar was the only district on the west coast often associated with GAM.

Pidie is well known as a "region of war", having experienced various "wars" since the 1800s, and suffered economically. Like Pidie, Bireuen was also one of GAM's strongholds where some of its sub-districts (*kecamatan*) were even called *daerah hitam* (black areas).

Administratively, there were only 13 districts with 11 regencies and 2 cities in 2000. However, in 2001, the government started to implement regional autonomy, to bring government closer to the people. The districts thus acquired much larger power which was in turn split into more districts. By 2005, there were 21 districts, consisting of 17 regencies and 4 cities (Sabang, Banda Aceh, Lhokseumawe, and Langsa) in NAD.

On the west coast of the province, the regency of West Aceh was split into three regencies: Aceh Jaya, West Aceh, and Nagan Raya. South Aceh was divided into Southwest Aceh and South Aceh. On the east coast, the regency of East Aceh was split into three districts, namely Aceh Tamiang, East Aceh and the city of Langsa. Other districts experiencing fragmentation were Central Aceh which was split into Bener Meriah and Central

FIGURE 2.1
Province of Nanggroe Aceh Darussalam by Districts, 2005

Aceh; and Southeast Aceh into Gayo Lues and Southeast Aceh.

Lhokseumawe became a separate district from the regency of North Aceh and was upgraded into a city. Lhokseumawe was considered an unsafe place because of the many political conflicts. The transnational oil and gas industry contributed most (86 per cent) of the city's gross regional domestic product in 2003. It was the richest district with per capita product of about US$4,877, five times the average per capita product of all districts in the province.[5] However, despite the fact that Lhokseumawe was the richest, 16.39 per cent of the population were living below the poverty line — compared to Banda Aceh with only 9.71 per cent, the lowest rate of poverty. (The poverty rate in Lhokseumawe was, nevertheless, much lower than the average of 29.76 per cent in the whole province of NAD.)[6]

Some elites from Aceh have proposed the formation of two separate provinces:

Photo courtesy of Lee Poh Onn

A mass grave for the victims of the earthquake and tsunami.

ALA (Aceh Leuser Antara) and ABAS (Aceh Barat Selatan), though Jakarta does not seem to be supportive of the idea. These elites declared the establishment of the two provinces in Jakarta on 4 December 2005, based on the "classic" argument that the centre (Banda Aceh) has never provided sufficient attention to the development of the two provinces. ALA, with its beautiful Mount Leuser National Park, would cover five districts in central Aceh: Bener Meriah, Central Aceh, Gayo Lues, Southeast Aceh, and Aceh Singkil. ABAS would include most of the districts on the west coast: Aceh Jaya, West Aceh, Nagan Raya, Southwest Aceh, South Aceh, and Simeulue.[7]

Bearing in mind the geography of these conflicts and also the impact of tsunami, the reconstruction, rehabilitation, and sustainable development planning of NAD cannot focus on the disaster area alone. Focusing on the west coast of the province would create social and political jealousy. Such a situation would hamper the effort to reconstruct, rehabilitate and develop the tsunami-hit areas. Special attention must be accorded to the district variations in terms of economy, values, politics, and natural resources. Because the reconstruction, rehabilitation and sustainable development should be conflict sensitive, the focus must be the number and composition of the population. The information includes the most basic aspects such as the size, age-sex composition, geographical distribution, religious and ethnic composition, as well as the number and distribution of the internally displaced persons.

This chapter examines the most basic characteristics of the population of NAD. The statistics are mostly from the latest available data set, collected by the Central Statistics Department of Indonesia (Badan Pusat Statistik) during the period 15 August–15 September 2005, eight months after the natural disaster. The survey was conducted especially to help in planning the reconstruction, rehabilitation, and development in the province. Information prior to the tsunami was drawn from the 2000 population census data set and is used to approximate the condition before the tsunami and to show a rough picture of what might have happened after the tsunami.

Age Structure of the Population[8]

The population of NAD was 4,031,589 in 2005. This number includes the 60,736 individuals without fixed places of residence or those who move from one place to another. Based on the population growth rates in NAD during 1980–2005, the number of tsunami/earthquake related deaths and out-migration is estimated to be between 250,000 and 350,000 people.[9]

As is a common trend in many other regions of Indonesia and in other countries, the population of NAD has undergone a transition from one characterized by high fertility and mortality towards one with low fertility and mortality. In 1971, the population of NAD had a very high dependency ratio, especially the young dependency ratio. The young dependency ratio was 85.5, meaning that for every 100 productive age (15–64) population, there were about 86 young, unproductive age (<15) population. The old dependency ratio was still low at 5.0, meaning that for every 100 productive age population, there were only 5 old, unproductive age (>65) persons. The total dependency ratio was 90.5, that is, every 100 productive age persons had to support about 91 unproductive persons. In other words, the society did not have much savings because of the high dependency ratio and hence they had little opportunity to invest.

The decline in fertility has resulted in a drastic decline in the young dependency ratio to 51.1 in 2005, meaning that in 2005 every 100 productive age persons had to support only 51 young unproductive age persons. On the other hand, the old dependency ratio increased to 5.8 in 2005, suggesting that in 2005 every 100 productive age persons had to support more elderly people. As a whole, however, the total dependency ratio was 57.0, much lower than 90.5 in 1971, implying that in 2000 the society could save much more and had more opportunity to invest than in 1971.

The ratio of 57.0 is relatively low. If the trend continues, the total dependency ratio in NAD will soon go below 50.0, when the burden of one unproductive person will be shared by at least two productive age persons. The age structure of a population becomes conducive for economic development — when a larger part of the income of the population can be saved and used for investment, rather for financing the unproductive population. This demographic opportunity should be optimally used to speed up the pace of sustainable development in the province.

Indeed, in 2005, some districts in NAD had passed the threshold, having a total dependency ratio smaller than 50.0. The population of the capital city of Banda Aceh was the most conducive for economic development, with the lowest total dependency ratio, of only 33.3. The next smallest total dependency ratio was found in the city of Sabang (47.4), another district hit hard by the tsunami. Others who already have the conducive age structure are the regencies of Aceh Jaya (49.8) and Aceh Barat (48.7). The regency of Aceh Besar just reached the threshold of 50.0. All these districts were located closest to the epicentre of the earthquake. Simeulue is the only tsunami hit district that does not have the conducive age structure. The least conducive age structure was seen in the regency of Aceh Singkil (75.9), in the western coast, followed by Aceh Tenggara (65.2) in the eastern coast.

In brief, this suggests that economic opportunities and development are greatest in Banda Aceh.

Geographical Distribution of the Population

After the tsunami, the population on the west coast of NAD was about one third of the total population in the province (see Table 2.1). The city of Banda Aceh is the capital of the province, with only 177,000 persons (or 4.41 per cent of the total population). Bordering with Banda Aceh is the regency of Aceh Besar,

TABLE 2.1
Number of Population by Sex and Districts: Province of Nanggroe Aceh Darussalam, 2005

District	*Male*	*Female*	*Total*	*%*
Western coast				
Banda Aceh	94,052	83,829	177,881	4.41
Sabang	14,663	13,934	28,597	0.71
Aceh Besar	152,377	144,164	296,541	7.36
Aceh Jaya	31,515	29,145	60,660	1.50
West Aceh	76,932	73,518	150,450	3.73
Nagan Raya	61,609	62,134	123,743	3.07
Southwest Aceh	56,809	58,867	115,676	2.87
South Aceh	93,684	97,855	191,539	4.75
Aceh Singkil	75,177	73,100	148,277	3.68
Simeuleu	40,519	37,870	78,389	1.94
			1,371,753	34.02
Others				
Pidie	228,404	245,955	474,359	11.77
Bireuen	169,767	182,068	351,835	8.73
North Aceh	241,942	251,728	493,670	12.25
Lhokseumawe	76,614	78,020	154,634	3.84
Central Aceh	81,016	79,533	160,549	3.98
Bener Meriah	53,168	52,980	106,148	2.63
Gayo Lues	35,488	36,557	72,045	1.79
Southeast Aceh	84,143	84,910	169,053	4.19
East Aceh	150,785	153,858	304,643	7.56
Langsa	68,518	69,068	137,586	3.41
Aceh Tamiang	118,581	116,733	235,314	5.84
			2,659,836	65.98
Total	2,005,763	2,025,826	4,031,589	100.00

Source: Compiled and calculated from Badan Pusat Statistik et al., *Penduduk dan Kependudukan Aceh Pasca Gempa & Tsunami. Hasil Sensus Penduduk Nanggroe Aceh Darussalam 2005* [Population in Aceh after Earthquake and Tsunami. Results of Population Census in Nanggroe Aceh Darussalam 2005] Table 2.2 (Jakarta: Badan Pusat Statistik, 2005).

accounting for 7.4 per cent of the population in NAD. What is significant is that the three largest concentrations of population were found in the conflict districts of North Aceh (12.25 per cent), Pidie (11.77 per cent), and Bireuen (8.73 per cent).

The distribution in terms of population density is also very uneven. The most densely populated areas were in the four cities of the province. Banda Aceh, one of the worst-hit tsunami areas, is the most densely populated area, with 2,916 persons per sq km. This is much larger than the one in the second most densely populated area, the city of Lhokseumawe, in the eastern coast of the province, with only 854 persons per sq km or more than one third as crowded as in Banda Aceh. The city of Langsa, also in the eastern coast, was much less crowded, with only 525 persons per sq km. The fourth was the city of Sabang, with a population density less than half of that in Langsa.

In other words, the population densities among the cities tended to vary, with that in the city of Banda Aceh about 28 times as crowded as the city of Sabang. Furthermore, the least densely populated area is the regency of Gayo Lues, with only 12.6 persons per sq km, a very huge contrast to that of Banda Aceh.

A Multi-Ethnic Muslim Province

The discussion in this section is based on the 2000 population census, to provide an approximate picture of the situation at the time of the tsunami. This information may provide policy-makers with an understanding of who may have suffered more.

The province of Nanggroe Aceh Darussalam is multi-ethnic and almost all of the population are Muslims. In 2000, Muslims formed 97.03 per cent of the population, with Christians, only 2.30 per cent. The Islamic identity among the Acehnese is very strong in their day-to-day life. Islam is deeply embedded in Acehnese history and culture. Interestingly, however, the strong identity with Islam has not been an integral aspect of GAM in that the proponents of GAM do not raise the issue of forming an Islamic state.

Nevertheless, one of the efforts of the Government of Indonesia to solve the conflict is to introduce the implementation of *shari'a* in the province by producing Law No. 44/1999. This law was strengthened with Law No. 18/2001, that granted special autonomy status to the province of Aceh special region and hence, since then, the province was named Nanggroe Aceh Darussalam. On the other hand, GAM and human rights watch groups believed that the implementation of the *shari'a* was not what the Acehnese wanted; the Acehnese had practised Islam in almost all aspects of their life for a long time.[10]

The province is not homogenous in its ethnic composition. The four largest "indigenous" ethnic groups in NAD are the Acehnese, Gayo, Alas, and Singkil. The Acehnese mostly resided on the west coast, the disaster area, while the Gayo and Alas were in the mountainous areas.

The major "outside" ethnic group is the Javanese, the second largest ethnic group in NAD, who lived outside the west coast. In other words, the Acehnese were more likely to have suffered from the earthquake/tsunami than the other ethnic groups in the province.

In 2000, only about half of the population in the province was Acehnese. As shown in Table 2.2, the Acehnese was a very small ethnic group in Indonesia. It formed less than 1 per cent of the whole population of Indonesia. Yet, overall, in 2000, about half of the Acehnese lived in the disaster area, on the west coast.

However, in some districts, the population was almost exclusively

Photo courtesy of Lee Poh Onn

A mosque in Meulaboh.

TABLE 2.2
Percentage of Population by Ethnic Groups: Province of Nanggroe Aceh Darussalam and Indonesia, 2000

Ethnic Group	*Nanggroe Aceh Darussalam*	*Indonesia*
Acehnese	50.32	0.43
Javanese	15.87	41.71
Gayo	11.46	0.10
Alas	3.89	0.03
Singkil	2.55	0.02
Simeulue	2.47	0.02
Batak	2.26	3.02
Minangkabau	1.09	2.72

Note: These are the six largest ethnic groups in the province of Nanggroe Aceh Darussalam.
Source: Compiled from Leo Suryadinata, Evi Nurvidya Arifin, and Aris Ananta, *Indonesia's Population. Ethnicity and Religion in a Changing Political Landscape,* Table 1.12 and Figure 5.4.1 (Singapore: Institute of Southeast Asian Studies, 2003).

Acehnese, with more than 95 per cent of the population being Acehnese. The largest concentration of the Acehnese was in the regency of Bireuen (96.88 per cent), the conflict area on the east coast, though the number of Acehnese in Bireuen only constituted a very tiny 1.20 per cent of the total Acehnese in the province. The second largest concentration (95.16 per cent) was in the regency of Aceh Besar, a disaster area in the west coast. It should be noted here that the 2000 population census could not record the population of the regency of Pidie because there was no guarantee for the security of the census-takers.

In addition to the regency of Bireuen, the former regency of North Aceh (which split into the current regency of North Aceh and the city of Lhokseumawe) was the only other district on the east coast with a very large percentage (89.61 per cent) of Acehnese. On the other hand, there were several districts in the disaster area (the west coast) with a very high percentage of Acehnese: the city of Banda Aceh, former regency of West Aceh (split into the current regencies of West Aceh, Aceh Jaya, and Nagan Raya), former regency of South Aceh (split into current regencies of South Aceh and Southwest Aceh), and the city of Sabang.

The second largest ethnic group in the province were the Javanese (15.87 per cent), the largest ethnic group in Indonesia as a whole. The Javanese constituted 41.71 per cent of the whole population of Indonesia, with Yogyakarta, Central Java, and East Java (all in the Island of Java) as their home provinces. Indonesian politics are heavily influenced by the Javanese, resulting in resentment outside the Island of Java against the Javanese and Jakarta (which is also located in Java). The large inflow of Javanese as military personnel and civil servants to NAD since the 1970s has also been seen by the locals as Javanese imperialism.

A very high percentage (80.47 per cent) of the Javanese did not live in the disaster areas. Specifically, two districts with the largest percentages of the Javanese were the former regency of East Aceh (32.86 per cent) and the former regency of Central Aceh (21.62 per cent). The regency of Aceh Singkil was the only district in the disaster area with a relatively large percentage (19.93 per cent) of the Javanese. Aceh Singkil was located in the southernmost part of the west coast, the farthest from the epicentre of the earthquake.

The third largest ethnic group was the Gayo, comprising both the Gayo Lut and Gayo Lues, forming 11.47 per cent of the population in the province, but only a very tiny percentage (0.10 per cent) in the whole Indonesian population. The province is also the home of the Gayo, with about two thirds of the Gayo Lut living in the regency of East Aceh and almost all of the Gayo Lues living in Central Aceh.

The fourth largest group, the Alas, was also an ethnic group whose home province is NAD. The Alas were often associated with the Gayo as both are originally from Southeast Aceh. The Alas formed 3.90 per cent of the population of the province, and lived mostly in Southeast Aceh. The fourth and fifth largest ethnic groups were also from the province. They were the Singkil (2.55 per cent) and Simeulue (2.47 per cent).

The remaining smaller ethnic groups were mostly from outside the province of NAD. Two of these groups were from the Island of Sumatra: the Batak (2.26 per cent), with North Sumatra as the home province, and the Minangkabau (1.09 per cent), with West Sumatra as the home province. Other groups were smaller, each comprising less than 1.0 per cent of the population of the province and they originated from outside the Island of Sumatra. They were the Sundanese from the Island of Java, Banjarese from the Island of Kalimantan, Bantenese from the Island of Java, Madurese from the Island of Madura, Buginese from the Island of Sulawesi, and Betawi from the Island of Java.

The Female Population

In 2005, the number of female population was about 20 thousand more than that of the male population.[11] Specifically, the female population outnumbered the male population among those aged 60 and over and those between 20 and 45. The excess number of female population aged 60 and over is not surprising because the population of NAD is already in the process of ageing with 6.08 per cent of the population aged at least 60 years. In almost all ageing populations, the number of old female population always exceeds that of the old male population. Declining fertility and continuing out-migration may have contributed to the ageing profile in NAD.

However, there were also excess numbers of female population in the young age groups. Is this because of out-migration of the young male population and/or the death of the young male population due to the internal war in NAD? To answer this question, the district level data must be examined,

Male workers at the damaged Muhammadiyah High School.

Photo courtesy of Mercy Relief

though the information on female and male population at the district level is not broken down into age-groups.

Although there was an excess number of female population for the whole province of NAD, there were deficits in several districts. Most of the districts were in the disaster areas. The city of Banda Aceh had the largest sex ratio (112.2), meaning that for every 100 female persons, there were about 112 male persons. The second largest sex ratio was 108.1 in the regency of Aceh Jaya, implying that for every 100 female persons in Aceh Jaya, there were about 108 male persons. Other districts with sex ratios larger than 105.0 were regencies of Simeulue (107.0) and Aceh Besar (105.7) as well as the city of Sabang (105.2). As seen from Figure 2.1, the city of Banda Aceh, regencies of Aceh Besar and Aceh Jaya were in the northernmost part of the west coast, closest to the epicentre of the earthquake. The city of Sabang and regency of Simeulue are small islands in the Indian Ocean, also on the west coast.

A deficit of female population was also found in three non-disaster regencies, but the deficit was very small. The largest one was only 101.9 in the regency of Aceh Tengah, meaning that for every 100 female persons in this regency there were about 102 male persons.

Therefore, the large deficit of female population in the natural disaster districts might indicate that the female population was more likely to have died during the tsunami-earthquake and/or to out-migrate from the districts after surviving from the disaster. However, the first explanation may be more plausible. A statistical study on pre- and after-tsunami clearly shows that the mortality rate was much higher among females than the male population.[12] Some possible explanations as to why women were less likely to survive during the disaster are discussed in Chapter IV of this book.

On the other hand, the excess of female population was obvious in the two districts in the conflict areas; the regencies of Pidie and Bireuen. In Pidie, the sex ratio was very low, 92.9, implying that for every 100 female persons, there were only about 93 male persons. The sex ratio in Bireuen was similarly low, 93.2. An excess of female population, though not as large as in Pidie and Bireuen, was also found in two other conflict areas, the regencies of North Aceh and East Aceh. The male population, especially the young, might have been more likely to die or out-migrate because of the internal war in these districts.

The regency of Aceh Besar was interesting because it was both a conflict and disaster area. As mentioned earlier, there was a large deficit of female population in this regency. This may imply that the disaster has affected the male-female composition more than the conflict has.

In most other districts outside the west coast, there was also an excess of female population, but the sex ratios were larger than 95.0. Interestingly, three districts in the west coast regions also experienced an excess of female population, but they are located far down in the south, further from the epicentre. They were the regencies of South Aceh (95.7), Aceh Barat Daya (96.5), and Nagan Raya (99.2). They might have suffered from the internal wars.

Internally Displaced Persons (IDP)

After the tsunami, at the time of the survey in August–September 2005, 5.21 per cent of the population of NAD were internally displaced persons, or IDP. 7.41 per cent had become IDP but were no longer IDP at the time of survey. In total, 12.62 per cent, or about 500,000, of the population had been IDP, and more than half were no longer IDP. Most of the

Photo courtesy of Mercy Relief

Temporary shelter.

IDP, especially those close to the epicentre, experienced the disaster personally.

As shown in Table 2.3, the percentage of the IDP at the time of survey looks small for the whole population of the province. However, the percentage was very large in the relatively poor regency of Aceh Jaya, one of the severely hit areas. Calang, the capital city of the regency, was completely destroyed. The IDP constituted almost half (45.76 per cent) of the population of the district. The regency actually has a beautiful landscape and is rich with natural resources, but the economy still depends on agriculture. In the early 1990s foreign visitors used to visit this regency, but they were later afraid to come because of the rising tension related to the "internal wars". Its per capita

TABLE 2.3
Numbers and Percentages of Internally Displaced Persons by Districts: Province of Nanggroe Aceh Darussalam, 2005

District	*Total*	*%*
Western Coast		
Banda Aceh	24,120	13.61
Sabang	2,061	7.21
Aceh Besar	46,998	15.85
Aceh Jaya	27,755	45.76
West Aceh	28,018	18.62
Nagan Raya	6,134	5.10
Southwest Aceh	2,146	1.86
South Aceh	4,547	2.37
Aceh Singkil	7,106	4.79
Simeuleu	15,498	19.77
Others		
Pidie	19,906	4.20
Bireuen	10,032	2.85
North Aceh	8,695	1.76
Lhokseumawe	2,476	1.60
Central Aceh	520	0.32
Bener Meriah	320	0.30
Gayo Lues	94	0.13
Southeast Aceh	304	0.18
East Aceh	1,411	0.46
Langsa	1,081	0.79
Aceh Tamiang	330	0.14
Total	209,822	5.20

Source: Compiled and calculated from Badan Pusat Statistik et al., *Penduduk dan Kependudukan Aceh Pasca Gempa & Tsunami. Hasil Sensus Penduduk Nanggroe Aceh Darussalam 2005* [Population in Aceh after Earthquake and Tsunami. Results of Population Census in Nanggroe Aceh Darussalam 2005] Table 3.1 (Jakarta: Badan Pusat Statistik, 2005).

product was only about 78.11 per cent of the average of the per capita product of the districts in NAD. Because the regency of Aceh Jaya had not exploited its rich natural resources, the per capita product in Aceh Jaya was even less than half the average of the districts after taking into account the contribution of oil and gas in other districts.[13]

In terms of the absolute number, nevertheless, the regency of Aceh Besar had the largest number of IDP: almost 47,000, though only constituting 15.85 per cent of the population of the regency. As mentioned earlier, both Aceh Jaya and Aceh Besar were close to the epicentre of the earthquake. Furthermore, all other districts not far from the epicentre of the earthquake had a large percentage of IDP: the regencies of Simeulue (19.77 per cent), West Aceh (18.62 per cent), Sabang (7.21 per cent) and city of Banda Aceh (13.61 per cent).

Interestingly, the number of male IDP far exceeds that of female IDP. In the whole province of NAD, there was about 117 male IDP for every 100 female IDP. This excess of male IDP was in contrast to the excess of female population for the whole population in NAD.

On the other hand, the excess of male IDP was particularly striking in the city of Banda Aceh and other districts close to the epicentre. The number of male IDP in Banda Aceh was almost 1.5 times that of female IDP: for every 100 female IDP, there were almost

Photo courtesy of Mercy Relief

Students keeping up with their studies and religious instruction.

150 male IDP. The next highest excess of male IDP was the regency of Aceh Besar, bordering with Banda Aceh. Interestingly, the regency of South Aceh, almost to the southernmost of the western coast, had the third largest excess of male IDP. The regencies of Aceh Jaya and West Aceh, close to the epicentre, had the fourth and fifth largest excess of male IDP.

Conclusion

The Acehnese lived mostly on the west coast — scene of the tsunami. The non-Acehnese, especially the Javanese, lived mainly in the conflict areas. Though there are no statistics on the victims of the natural disaster by ethnicity, the geographic picture of the location of ethnic groups and the events of the natural disaster may provide a hint that the natural disaster struck the Acehnese more than the other ethnic groups in the province of NAD. On the other hand, the Javanese, the largest ethnic group from outside the province were less likely to have suffered from the disaster. The disaster may therefore have changed the ethnic composition in favour of the non-Acehnese, especially the Javanese.

The multi-ethnicity factor, especially between the Acehnese and other "indigenous" ethnic groups on one hand and the Javanese on the other, should warrant special attention, given the perceived hatred towards the Javanese and the fact that the Acehnese were more likely to have suffered from the natural disaster than the Javanese. The large regional (district) variation also warrants careful consideration to maintain the peace momentum created by the Government of Indonesia and GAM.

In other words, though the impact of the natural disaster was felt more in the west coast, the programme for reconstruction, rehabilitation, and development in the province must not be at the expense of those in the conflict areas, beyond the west coast. Yet, the population in the conflict areas had not received as much aid as the disaster-hit areas had, despite the fact that the conflicts had caused long suffering and

Photo courtesy of Lee Poh Onn

Reception at an orphanage during celebrations for Indonesian Independence Day, 17 August 2006.

the province might have experienced a larger degree of casualty and economic destruction.[14] With less aid to the conflict areas, the long conflict can be easily re-ignited. Such conflict would again slow down the process of sustainable development in the province.

One of the major problems is that the BRR NAD-Nias, which is assigned by the Government of Indonesia to carry out the reconstruction, rehabilitation, and development of the tsunami-hit areas, does not have the authority to work on post-conflict districts. The programmes for the post-conflict districts are still separately managed by the governor's office and central government.[15]

However, with wise planning, accommodating the ethnic and regional diversity of the province, the province will be able to smoothly reconstruct and rehabilitate; and, subsequently, to experience sustainable development.

Notes

1. The author acknowledges the valuable inputs and comments from Evi Nurvidya Arifin, Tiodora Siagan, and Siswa Rizali.
2. Calculated from Badan Pusat Statistik, *Gross Regional Domestic Product of Provinces in Indonesia by Industrial Origin 2002–2003*, Tables 3 and 4 (Jakarta: Badan Pusat Statistik, 2004).
3. The loss on the gross regional domestic product was estimated by the World Bank, as cited in BRR NAD-Nias, *Aceh and Nias One Year after the Tsunami. The Recovery Effort and Way Forward*, Map 4.1 (Banda Aceh, Indonesia: BRR-NAD-Nias, 2005).
4. Badan Pusat Statistik, *Gross Regional Domestic Product of Provinces in Indonesia by Industrial Origin 2002–2003*, Tables 9 and 10 (Jakarta: Badan Pusat Statistik, 2004).
5. Calculated from Badan Pusat Statistik, *Gross Regional Domestic Product. Regencies/ Municipalities in Indonesia. 2000–2003*, Table 121 (Jakarta: Badan Pusat Statistik, 2004). The exchange rate is assumed to be 9,000 rupiah per one U.S. dollar.
6. Badan Pusat Statistik, *Data dan Informasi Kemiskinan Tahun 2003. Buku 2: Kabupaten* [Data and Information on Poverty 2003. Book 2: Districts] (Jakarta: Badan Pusat Statistik, 2003).
7. *Kompas*, "11 Kabupaten di Aceh Tuntut Provinsi Baru. Partai politik lokal harus bisa ikut pemilu nasional" [11 Districts demand new provinces. Local political parties must be allowed to participate in national election], *Kompas*, 12 October 2005, and Sidik Pramono and Ahmad Arif, "Jika Aceh Manjadi Tiga" [If Aceh Becomes Three], *Kompas*, 14 August 2006.
8. The statistics for the population in 2005 are compiled and calculated from Badan Pusat Statistik, *Penduduk dan Kependudukan Aceh Pasca Gempa & Tsunami. Hasil Sensus Penduduk Nanggroe Aceh Darussalam 2005* [Post-Earthquake and Tsunami Population of Aceh. Results of the 2005 Population Census. Nanggroe Aceh Darussalam.] (Jakarta: Badan Pusat Statistik, 2005).
9. The estimation is carried out using the following steps. First, the annual population growth rate in 1990–2000 is assumed to continue in 2000–05. Then, this growth rate is used to calculate the number of population in 2005 if there had been no tsunami/earthquake. Because the actual population growth rate tends to decline, the assumed growth rate in 2000–05 is the maximum

one. Therefore, the difference (loss) of population, calculated by comparing the estimated and the "real" number of population in 2005, is the maximum number of tsunami-earthquake related deaths and out-migration. Second, the annual population growth rate is assumed to decline to a maximum 1.75 per cent in 2000–05. With this rate, the minimum number of tsunami-earthquake related deaths and out-migration is estimated.

10. Andreas Harsono, "Struggle Over Sharia" in *The Irrawadi*, On-line Edition, 1 October 2003.
11. The statistics are compiled and calculated from Badan Pusat Statistik, *Penduduk dan Kependudukan Aceh Pasca Gempa & Tsunami. Hasil Sensus Penduduk Nanggroe Aceh Darussalam 2005.* [Post-Earthquake and Tsunami Population of Aceh. Results of the 2005 Population Census. Nanggroe Aceh Darussalam.] (Jakarta: Badan Pusat Statistik, 2005).
12. Elizabeth Frankenberg et al., "The Impact of a Disaster on Life in Sumatra: Evidence from before and after the December 26 Tsunami". Paper presented at the annual meeting of the Population Association of America, New York, 30 March–1 April 2006.
13. The statistics on gross regional domestic product is calculated from Badan Pusat Statistik, *Gross Regional Domestic Product. Regencies/Municipalities in Indonesia. 2000–2003*, Table 121 (Jakarta: Badan Pusat Statistik, 2004).
14. BRR NAD-Nias, *Aceh and Nias One Year after the Tsunami. The Recovery Effort and Way Forward* (Banda Aceh, Indonesia: BRR-NAD-Nias, 2005).
15. Ibid.

Photo courtesy of Lee Poh Onn

House built by Habitat for Humanity, Singapore.

3

Recovery and Reconstruction

Kuntoro Mangkusubroto[1]

When the tsunami struck in December 2004, many areas were devastated. This was especially so in the western parts of Aceh where the port and inland roads were completely destroyed. The only way to transport medical facilities and food was by air. This was initially undertaken by the Chinook helicopters provided by Singapore. Since then, several Singapore NGOs like Mercy Relief, the Singapore International Foundation (SIF) and the Singapore Red Cross have been helping Indonesian tsunami victims, especially in the reconstruction of the Meulaboh harbour. This habour is now complete and became operational in April 2006, serving as one of the most strategic points of entry for building materials and supplies to the area.[2]

This chapter is an update of the rehabilitation and reconstruction programme, detailing the progress taking place throughout Aceh and examining many aspects of the reconstruction and rehabilitation process. It should be noted that the agency — BRR, the Badan Rehabilitasi dan Rekonstruksi of Aceh and Nias — was formed by the Government of Indonesia on 30 April 2005. Two days after that, on 2 May, the BRR was mobilized without delay and its forces were put into operation in Banda Aceh. Since then, its operations and base have been in Banda Aceh. The BRR has over 700 full-time

Facing page: Coastline of Meulaboh. *Photo courtesy of Lee Poh Onn.*

personnel, mostly resident in Banda Aceh, some bringing along their families to settle in the area.

The formation of the BRR was based on a special bill: *Peraturan Pemerintah Pengganti Undang Undang* No. 2/2005, which was later formalized into the *Undang Undang* No. 10/2005.[3] The BRR is to operate for four years, and thereafter it will be handed over to the local government and technical ministers in Jakarta to continue with the process. Within four years, the management at the BRR hopes that all major infrastructure projects will be completed, laying the foundation for further development in the area.

Three big missions lie ahead for the BRR. One is to rebuild all facilities, infrastructure, and houses within four years. The second is to improve the capacity of the local government and social institutions. The third is to revitalize businesses and the economy in the region such that the sustainability of the whole programme can be maintained.

Progress at a Glance

Figure 3.1 shows a high-school complex specializing in agriculture at Kahju, Aceh Besar. It is about 40 kilometres from the totally destroyed Banda Aceh. It is in the process of being reconstructed. In total, over 375 schools have been rebuilt, representing a significant achievement. It should be noted, however, that 1,200 schools were destroyed during the tsunami, including elementary and high schools and sections of a university. Hence the proportion of reconstructed schools only represents about 35 per cent of the total number of schools. Areas in Aceh now look relatively clean. A year earlier, it was still covered with debris. Much progress has been achieved in the previous one and half years.

Figure 3.2 illustrates a temporary housing project, also in Aceh Besar. There are various types of housing in the vicinity. BRR does not standardize their design but stipulates the size of a house. Previously, the specifications were such that it should be no less than 45 square metres. Then, BRR realized that they were going to face problems if they did not put an upper limit to the size of the houses because some NGOs in the area were competing with each other to provide better and bigger housing.

Although one NGO may be willing to build 45 square metre housing in some areas, some 70 square metre houses can still be found elsewhere. This causes a social problem when villagers start to compare the size of housing provided: why in one village is there 70 square metre housing, while in others 76 square metre housing is available? The 45 square metre specification has been selected because in some areas, especially in Meulaboh, many houses have already been built based on that size limit. Basically, these are known as core houses. Owners of such houses can expand and redesign from this core. The basic design of these houses is such that these houses are ready to be "expanded", if owners chose to exercise the option of doing so.

FIGURE 3.1 Educational Complex at Kahju, Aceh Besar

BRR targeted some 30,000 houses to be built in 2005. By February 2006, it had built 34,000 houses, a little over the target. There is still a long way to go, however. It is hoped that by mid-2007, all 127,000 houses will have been built.

So far, about 70 per cent of the needed housing facilities have been constructed, and 41 per cent of the hospitals and clinics completed. In terms of educational infrastructure, 675 schools have been reconstructed, representing around 16 per cent of the total, as the total number of schools destroyed was in the thousands. Sadly, there are more than 2,500 cases of deceased or missing teachers. It is an enormous challenge to

FIGURE 3.2 Temporary Housing in Krueng Raya, Aceh Besar

replace 2,500 school teachers in a short space of time.

For small-to-medium enterprises, the BRR has a micro-financing scheme for all new activities. With respect to infrastructure, only over 7.8 per cent of the roads and five major harbours have been restored. Only two harbours, however, are now operational. The major port, Malayahati, was completed in August 2006.

Figure 3.3 illustrates the extent of the fish/shrimp ponds that were destroyed or severely damaged. The total area damaged was more than 800 kilometres of the coastline. The fish/shrimp ponds were among the major areas destroyed. Out of the 60,000

FIGURE 3.3 Rehabilitated Fish Pond in Krueng Raya, Aceh Besar

hectares of shrimp/fish ponds, 25 per cent have been cleaned of the inordinate amount of debris and tree stumps.

Cleaning shrimp ponds has been a big project. In some areas, for instance in Bireuen, where the areas of damaged shrimp ponds stretched over 12,000 hectares, around 6,000 hectares have now been cleaned. Not all shrimp ponds, however, can be cleaned easily, because apart from the debris, sea mud has also engulfed and polluted these ponds. Chemicals need to be used to restore these ponds so that they can be worked once more. Nonetheless, the 25 per cent already cleaned and operational represents a significant accomplishment. Of the 6,000 hectares already cleared, 2,500 hectares are now available for villagers to start breeding shrimp again.

The city of Banda Aceh is very close to the sea, and parts of the city are still covered in water. Banda Aceh is basically a beautiful city because there are four river systems crossing through it. A wave breaker needs to be constructed to keep the area dry. This is a major project as BRR needs to construct a 23 kilometres wave breaker. Because of the way the river system runs, some dykes also need to be built. A total of around 16 kilometres of dykes are needed as a part of this process of reconstruction. For a wave breaker, gates are not needed, but for a dyke, gates are needed to control the water flow. Between the dykes and the wave breaker, fish ponds and shrimp ponds can be built. Certain areas need to be dried before houses can be built. Unless these dykes and wave breakers are completed, nothing can be done in these areas. The construction of the wave breaker in Banda Aceh is expected to be completed in April 2007 and the dykes by December 2006. Thereafter, the drying of the area and the cleaning of these fish ponds and shrimp ponds can be started.[4]

The only cement factory in Aceh Besar is under reconstruction. It is currently under repair and, upon completion, will commence production. This is a very important facility, as cement is a major component in the reconstruction process.

Figure 3.4 shows the highway connecting Banda Aceh to Meulaboh. The process of reconstruction has already begun with around 10 per cent currently completed. Upon completion, which will take approximately another two years, it will have a distance of between 200 to 240 kilometres. As a precaution, this road has been built, where possible, at a further distance from the ocean. The previous road was too close to the sea and was therefore completely destroyed when the tsunami struck.

A temporary road has already been built and currently is the major backbone for transporting building materials. The temporary road, however, is relatively close to the sea and for this reason remains vulnerable. Constant repairs have to be carried out on this road because of the continual use by trucks carrying materials used in construction. As such, many activities are carried out just to repair the road and to keep it in working condition. Frequent road repair activities can be seen along the way. Until the main highway is completed, people will have to tolerate this condition.

Calang is the most devastated area in western Aceh. Before the tsunami, there were 80,000 people living in this area. After the tsunami, only 25,000 survived. A significant number of activities are underway in this area as indicated in Figure 3.5.

Figure 3.6 shows the pier at Meulaboh Harbour. It is a 175 metres long pier, with a depth of about 12 metres. With this pier, people do not have to rely on the road infrastructure alone. Cement, timber and tiles can now be transported via the pier.[5] This is the

FIGURE 3.4 **Banda Aceh-Meulaboh Road Project**

only operational pier along the west coast of Aceh. Without the pier, logistics were difficult for BRR. The rebuilding of this pier has been assisted and financed by the Singapore Red Cross through funds donated by the people and the Government of Singapore.

The city of Singkil is seldom mentioned in the press. Parts of the city have sunk by about 30 centimetres and other parts by as much as one and a half metres. One third of the city of Singkil was covered in water, and the port is no longer accessible because it has been submerged. Conversely, the entire Simeulue Island has been raised by about one-and-a-half metres. As such, the completion of the pier at Meulaboh is

FIGURE 3.5 Calang

June 2005

November 2005

FIGURE 3.6 Reconstruction of Meulaboh Port

very strategic for the transportation of all building materials to reconstruct this area. The BRR is responsible for rehabilitating and reconstructing the damaged areas along the seashore and also the islands, including Simeulue and Nias.

At least 120,000 houses need to be rebuilt. Ten per cent of them have to be relocated because the land is now covered by water and there is no way to build new houses on such areas. Over 70,000 houses were damaged, and they require reconstruction. Fourteen seaports have been destroyed. One hundred and twenty bridges were damaged.[6] From these statistics, the devastation caused by the tsunami across vast areas of Aceh becomes clear.

It is worth mentioning that 60,000 hectares of agricultural land had been damaged and the 100,000 small and medium entrepreneurs lost their livelihood. This is something that has significantly affected the economy of Aceh. It is therefore not surprising that the third mission of BRR is to revitalize businesses and the economy in Aceh. The reconstruction of infrastructure needs to be accompanied by economic revival; otherwise the facilities that have been built by BRR would not result in longer term and sustainable economic development in the area.

The progress in Nias, which is located off the coast of North Sumatra, has been a little slower but it is now catching up.

Countries around the world should be credited for providing assistance to the people of Aceh. The BRR is not only an agency directly accountable to Jakarta but accountable to donors, the NGOs that are helping the BRR, and also the lay people around the world who have donated cash and in kind to Aceh.[7]

Altogether, a pledge of US$7.1 billion had been made as of April 2005. The world has placed great trust in the BRR, such that US$4.5 billion has been committed to the BRR. Out of that US$4.5 billion, around US$2.5 has been allocated in the form of contracts or disbursements.

Hence, the US$4.5 billion given to an agency like the BRR that was formed only ten months earlier represents a major achievement. Out of this total, the various NGOs have contributed US$982 million. This is an unprecedented figure. If the International Red Cross is included among the group of NGOs, then the figure comes close to or a little more than US$1.5 billion. Direct bilateral donations add up to US$1.4 billion while those directed through the multilateral donations, such as the Multi-Donor Fund, the World Bank, and ADB add up to US$1.2 billion. This Multi-Donor Fund is US$450 million; the United Nations, US$391 million; and the Government of Indonesia, US$2.1 billion.

Funding Priorities

The first priority is housing. BRR plans to complete construction of the 120,000 houses destroyed, and the rehabilitation of the 70,000 damaged houses by mid-2007. This represents an extremely tall order. The total capacity of Indonesia to build public houses is not more than 70,000 houses a year. In this instance, the BRR is going to build about 78,000 houses in one year which is greater than the present national capacity. And by 2007, BRR will complete the remaining housing needs. One method includes the use of pre-fabricated houses imported from Bangkok, Kuala Lumpur, and Canada.

Why does the BRR have to finish this project by the middle of 2007? Because the BRR wants to ensure that people do not live in barracks, tents or temporary shelters for more than two years. It is very difficult to live in such conditions. After the houses have been built, BRR hopes to begin with the reconstruction of the support infrastructure and other activities.

The year 2009 will serve as a milestone, because the BRR will end its commission in that year; it may exist in another form, but according to the bill, BRR has to pull out from Aceh in 2009.

The big master plan for Aceh is that all areas will be connected to Banda Aceh served by three major roads. The Sabang Island Harbour will be the major hub from Singkil to Banda Aceh, from Kutacane to Banda Aceh, and from Kuala Simpang to Banda Aceh. The BRR wants to make Sabang the major harbour — the main hub for all commodities and economic activities in the hinterland of Aceh. Can this be developed further? Can it become a major hub in this region or, if not, are there any other possibilities like the Lhokseumawe Harbour?[8]

The United States and Japan are working together on Meulaboh. The U.S. component alone is currently the largest infrastructure development in Indonesia. Singapore has helped to build the harbour in Meulaboh. Mercy Relief, SIF, and Singapore Red Cross have been concentrating their efforts in Meulaboh. They have been helping Indonesians in the area because, in part, the Singaporean Armed Forces were the first to arrive to help the people of Meulaboh, which subsequently resulted in this close emotional bond.[9]

Community-based Development

Decisions on the reconstruction and redevelopment of villages need to made by the community. Over 600 villages were destroyed. Family members were killed, their houses were ruined as were documents, land certificates, their diplomas; in short, everything was destroyed. How can a village be rebuilt? From a legal standpoint, there is no way the whole process can even begin because of the difficulty of ascertaining land ownership. So the pragmatic approach has been to let the people decide, to let them draw up the plans, allow them to know that the future is in their hands, and that BRR is there only as a facilitator.

In relation to village planning, it is the people in the village who actually make the plans. An example would be the village in Lamjabat which was totally destroyed. BRR allowed the villagers to discuss, hold meetings, and then decide how their village is to be rebuilt. The BRR provided overall guidance, warning about dead-end streets, that should another major crisis like this occur again, such a road could become a killing field. Thus, the BRR advised villagers against having any dead-end streets in their village. The BRR also provided advice on the width of roads (village streets should not be narrower than four metres, ideally around six metres). By providing such guidelines, the BRR is hoping that the deadly impact of future natural disasters will be averted or reduced.

The BRR suggested that mosques be constructed in such a way that the upper level can become another safety area for emergencies. The location of amenities like ponds, schools and the village clinic has also been based on a consultative process with the community. In terms of land ownership, the same consultative process has been used. After the group has come up with a consensus of who owns which plot(s) of land, their signatures are put onto the planning map, and the map basically becomes an official document.

Once consensus and signatures have been put on the map, BRR looks towards the NGOs who have expressed an interest in building these houses. If there is no NGO offering to build in a particular area, BRR then takes the responsibility of building the houses. This is essentially the approach that BRR has been adopting.

As Aceh had been in conflict for thirty years, rebuilding a community and village is not an easy task. It is especially challenging for an outside body like the BRR to come to the area and to assist in rebuilding lives and also villages in the vicinity. Establishing working relationships takes time and a great deal of effort on all sides.

The process of rebuilding may also take some getting used to. Many activities happening in an area may inadvertently lead outsiders to conclude that the reconstruction process in a village is near completion. The process is on-going, however, and it will take some time before the construction process can be completed.

At the sub-district level, it is the integration of one village with another. For example, villagers complained about dead-end roads. When the BRR entered another village, the villagers there did not want to give their piece of land for the continuance of a road to enter into their village. BRR therefore requires a consensus at the sub-district (*kecamatan*) or the district (*kabupaten*) level. By doing this, step by step, decisions can be made based on a consensus starting from the micro level up to the macro where BRR reaches to the *kabupaten* or provincial level.

At the *kabupaten* level, it can be difficult to get a consensus to start to build hospitals, an airport, a harbour, or a power station. But once this step by step consensus approach has been used, then the integration of the whole province can be much easier. This bottom-up approach is new in Aceh. It involves slow, painstaking work that builds a community as well as physical buildings. We follow it wherever possible but recognize its limitations in some circumstances where the need for speed outweighs other considerations.

The BRR is a non-political and non-sectarian agency. It is a professional agency which has proven itself as the leader of the reconstruction programme despite the tremendous challenges such a programme faces. One such area is governance. The BRR has supported major reform in this area. What was historically a tense and chaotic region has changed almost overnight. It is difficult to believe, for instance, that the peace process could take place in such a short time and with such amazing results. Aceh will have a new life.[10]

A combination of factors have contributed to this change. The sincerity and efforts of the BRR is paramount: it is there to help, to rehabilitate and to reconstruct; it has provided the people with the hope of a new beginning. The visible and beneficial activities taking place in the province attest to BRR's efforts. The Acehnese sincerely desire change. They want greater transparency, and less corruption and human rights abuses.

Anti-corruption Drive

As noted earlier, about US$7.1 billion has been pledged from all over the world. School children like Maggie in Milwaukee sent her lunch money that she collected for a week to Aceh. A retiree in Paris has sent $2 to Aceh. Big governments have donated US$1 billion to the Acehnese. Regardless of the size of the donations, individuals and governments from all over the world do not want such donations to be misused. As such, right from the beginning, governance and accountability have been important priorities and are on the top of the list for the Indonesian Government.

At the onset, BRR set up — right from day one — all instruments, units, and monitoring mechanisms such that BRR, its activities, and all related parties in the reconstruction of Aceh will not be tainted by corruption. The BRR imposed a code of ethics and an external anti-corruption declaration. Everybody related to the reconstruction process had to sign this declaration.

Corruption is not an easy aspect to discuss and it is well known that the previous Governor of Aceh is in jail for corruption. In fact, corporate corruption is everywhere. Bribery is everywhere. The BRR is actually a small island of integrity in a sea of such practices. But even so, we have our challenges and we must be ever vigilant in our efforts.

If all activities are to be "clean", then the role of local governments cannot be ignored, especially in the process of capacity building in dealing with corruption.[11] Developing a complaints handling system and utilizing an investigation system is crucial — the BRR is the only government institution with an anti-corruption unit. This anti-corruption unit is advised by Betrand De Speville, who is one of the commissioners of the legendary Hong Kong Anti-Corruption Commission in the 1980s. Bertrand is with BRR to advise on what can be implemented and to maximize transparency through an IT-enabled solution. These are some of the things that the BRR has laid down in its anti-corruption strategy. All activities in Aceh are transparent. When it comes to finances, the public can view the BRR's reports.

The BRR has a system such that all its activities are in its database. This is known as "The Recovery of Aceh and Nias Database" (RAND). Basically, it is the Aceh's DAD.[12] Data are inputted from all activities around this area by NGOs, donor countries, or other project proponents who submit their project concept notes to the BRR.[13]

Therefore, individuals or entities who would like to be informed of the progress in Aceh can do so by accessing our website at <*www.e-aceh-nias.org*> and then enter the database by selecting the sub-menu RAND. If people want to see the progress made by Mercy Relief, for example, and their activities in Aceh, this website contains such information. For example, what is happening in housing can be obtained from the database on housing projects, and also the NGOs who are helping in building these houses. To gather information on Habitat for Humanity, their link on the site provides such information. Data can be obtained, for example, on the number of houses that have already been built, and the numbers that are yet to be built. By having this database, the headquarters and home base of the respective NGOs do not need to wait for a hard copy report from Aceh. They can see, and they can monitor progress in Aceh and also their own NGOs through the website.[14]

There were around two hundred and ninety NGOs working in Aceh. Coordinating the activities and preventing overlaps with such a large number of NGOs has indeed been a challenge. To resolve this matter, there is, for example, a map of Aceh on the BRR's website. In the map of Aceh, the user can zoom in on any *kabupaten* (district) or *kecamatan* (sub-district). Further, in *kecamatan* Krueng Sabee, as an example, even a *desa* (village) can be located. In this data base, there is information on how many houses have already been built, the number of houses required in the various *kabupaten*, *kecamatan*, and even *desa*.

The problem sometimes lies in inputting the data when there is an absence of Wi-Fi connection in some areas. In that case, a hard copy must first be produced. The hard copy is then sent to the district or *kabupaten* capital city, and hopefully there is a Wi-Fi connection there. Sometimes NGO and donor staff have been seriously challenged with this process. But in areas with automatic connection to Wi-Fi, the data can be inputted immediately. By doing this, the progress of a project on housing, fish ponds, hospitals, and schools can actually be monitored. Information is also provided on who is building houses in an area and who is building schools in the same area. With this system, the potential for overlaps in certain areas can be identified and resolved. If there is an NGO coming to Aceh with the intention to donate, say, to an orphanage, this database can be opened and they can locate the orphanage guided by the data in the map.

Up to 2005, the donation process in Aceh had been a supply driven approach: efforts were sometimes *ad hoc* and in certain areas there were more supplies than needed. BRR wants to reverse this anomaly and base donations on demand. Using this approach, the BRR can direct

NGOs or donor countries on where to donate their specific donations. This is already on the website.[15]

Widespread support is needed by the BRR for its rehabilitation and reconstruction efforts. There is a long way to go. The BRR is so glad that the peace process has now actually been achieved such that the reconstruction and rehabilitation benefit not only those affected by the tsunami but also those affected by the conflict. As mentioned earlier, a strong commitment lies in matchmaking supply and demand. Previously there was more supply than demand. If the BRR cannot connect supply to demand, then the BRR cannot coordinate activities efficiently. With this system, the BRR can match needs with supplies, such that there is no oversupply in certain areas, with a common concern to promote good governance.[16]

Every party wants to know how the money is being spent — when, how, the quality, and so forth. Therefore, by having good governance, basically the BRR is held accountable, and accountability is present in all aspects of its operations.

Out of the US$7 billion promised to Aceh, the BRR manages only those funds channelled through the BRR. In the case of the pier that was constructed at Meulaboh, the BRR left the process of governance and procurement procedures to the Singapore Government. Literally the BRR had to say: "Please you have your own governance. You have your own procurement procedures. Feel free to implement all the procedures to build that harbour. We are grateful that at the end of the day, this harbour will be provided, will be finished and be operated for the sake of the development of the people of Aceh".[17]

Thus, when it comes to money from the outside, NGOs or government institutions, the BRR does not interfere with their procedures. If they want to use their own consultants, their own contractors, their own materials, or even their own labourers — they are free to exercise that option. In Banda Aceh, there were sixty elementary school buildings and pre-fabricated buildings under construction with some of the labourers even coming from China.

In order to facilitate the whole process, the BRR has established a one-stop unit to facilitate the extension of visas and clearance of tax-exempted materials coming into Aceh from abroad. Through this unit, the BRR processes the flow of foreign materials from the Belawan Harbour if it is in Medan, Malayahati Harbour or some other harbours in Aceh through its customs processing.

Concluding Remarks: Recovery Aceh Nias Trust Fund

Donor countries and multi lateral institutions had set up a Multi-Donor Fund and some funds have been passing through this channel, mainly for the bigger players. At that time, requirements were such that only entities which donated or granted more than US$10 million could go through this channel. There were many smaller institutions that actually wanted to donate funds, but had not been able to do so. These smaller institutions had requested the BRR to set up another trust fund, the "Recovery Aceh Nias Trust Fund", for the non-traditional donors and private foundations like British Petroleum, the BNI or other banks, Alcatel, and other smaller establishments.

Some have sent US$200,000 through this trust fund, which is managed by BRR in partnership with HSBC, Standard Chartered Bank, BNI, and Deutsche Bank. The BRR wants to ensure that everything is professionally managed in a very transparent way. Through this trust fund, the available funds can be directly used to support projects.

For instance, if an institution in Singapore wants to donate, say, S$200,000 and the entity specifically mentions that the money is to be used to rebuild an orphanage, but it does not have any other institution in Aceh to carry out this request, then the institution in Singapore can donate through the trust fund and the BRR will manage the donation such that the request can be achieved.[18]

Finally, the BRR has been working hard for the people of Aceh and Nias and it wants to ensure that all donations coming from donors all over the world will be spent efficiently, wisely, and directly for the benefit of the people of Aceh and Nias.

Notes

1. This chapter is based on an ISEAS public lecture by Dr Kuntoro Mangkusubroto, Director of the BRR (Rehabilitation and Reconstruction Agency) on 2 March 2006.
2. The pier was completed and became operational on 3 April 2006.
3. *Peraturan Pemerintah Pengganti Undang Undang* or *Perpu* is Regulation in Lieu of a Law; *Undang Undang* is Law.
4. At the time this book was prepared, the wave breakers and dykes were already functional.
5. Ibid.
6. The remaining statistics are not mentioned here because it would be too long.
7. In the month of February 2005, my wife and I took a walk along the Singapore River and saw school children with a small donation box for the Tsunami Aceh. I have to be accountable to the school children who have collected donations for us, too.
8. I visited the Port of Singapore Authority (PSA) to get their assessment on the viability of Sabang Island as a port.
9. When I mentioned this to President Yudhoyono, he suggested that I ask Singapore to help us in Meulaboh because emotionally they are attached to that area.
10. I live in Banda Aceh. I moved my wife there ten months ago so I am close and have close contact with individuals from all walks of life. I can tell you that this peace process will last.
11. I have had discussions at the Lee Kuan Yew School of Public Policy and also the Civil Service College with regard to capacity-building for the local government officials of Aceh.

12. Development Assistance Database (DAD) has been used in other tsunami-affected countries such as Thailand, the Maldives, and Sri Lanka, with the main purpose to strengthen the effectiveness and transparency of the coordination of the use of the funding. <www.tsunamispecialenvoy.org/printable/transparancy.asp>.
13. The RAND system is actually a managerial information system to monitor the implementation of the budget and to ensure that there is no overlap of projects. <//e-aceh-nias.org/brr_oneyar/BuildingLandofHope/4Chapter3.pdf>.
14. There was an incident when a German NGO reported to his headquarters in Germany that they had already built one hundred and fifty houses in Pulau Aceh, in Aceh Island. Actually they had only built around fifteen houses. When a report seeking verification was submitted to BRR, I told them "Why don't you just open our website and you can see the link to this NGO?".
15. Kuntoro Mangkusubroto hoped that the readers will access the website and find the needed information. If information is not available, the BRR can be contacted directly and they will also check why the information cannot be accessed.
16. In the *Sydney Morning Herald*, there was an article criticizing the Australian support for Aceh because of lack of transparency. Actually, this was Australia's own problem.
17. As mentioned earlier, the pier was then completed on 3 April 2006 when it became operational.
18. A brochure on the trust fund can be directly requested from the BRR.

4

The Women of Aceh

Braema Mathiaparanam

As described in Chapter II, the province of Nanggroe Aceh Darussalam (NAD) had a larger female than male population even after the tsunami. This is partly because many men among the Acehnese had either migrated or died in the ongoing conflict of the Free Aceh Movement insurgency and partly because naturally women have longer expectancy of life. Women were left to carry on as breadwinners and caregivers. During the earthquake and tsunami that struck along the 800 kilometre coastline of the province on 26 December 2004, more women perished because there were more of them there at that time and also because, as will be elaborated later, they were simply unable to escape the disaster.

This has contributed to a gender imbalance in Aceh. But it is the figures at the village level that localize the impact of the devastation and shift us beyond the usual big death toll figures. For example, in the village of Deyah Mapplam, only 270 survived among the 4,500 villagers. Of those survivors, only a third, or 90, were women.[1] Flower Aceh[2] was among the earliest non-governmental organizations in Aceh which compiled

Fewer women than men survived the tsunami in Aceh. Of those who survived, many are now single parents, having lost their husbands to the tsunami or even earlier, to the military conflict. *Photo courtesy of Mercy Relief.*

Soldiers in a mountainous area.

Photo courtesy of Mercy Relief

Women taking on responsibilities in the rehabilitation and resettlement programmes.

Photo courtesy of Mercy Relief

reports on five villages in Aceh's Lampu'uk sub-district. They showed that only 40 out of 750 survivors were women in an area where the population was once 5,500.[3] Another report, Oxfam International's, on North Aceh showed that the death toll among the women from four villages was 70 per cent of all fatalities. Out of the 676 persons that was recorded, there were only 189 women survivors; male outnumbered female survivors by a ratio of 3:1.[4]

Many reasons have been offered as possibilities to understand this gender imbalance among the survivors. The women might have been more vulnerable as most were not good swimmers; they might have lost time and energy in struggling against the waves with children and the elderly in tow, and were hindered from climbing trees or from running fast because of their traditional outfits. The second possibility is that, even before the tsunami, women already outnumbered men as many men had either fled Aceh or had been killed in the conflict between the Indonesian military and the GAM (Free Aceh Movement). In fact, reports show there were already about 148,000 women who were single parents who lost their husbands in the armed conflict or whose husbands had become either internal or external migrants.[5] Under such circumstances these women were both the primary caregivers and breadwinners. The third possibility is that the men were more likely to be outside the houses during the disaster and hence they had more chances to save themselves. Even now there is no clear evidence to show which factor or factors contributed to this gender imbalance at the district level.

Impact on Women

The tsunami also robbed women of their home industry livelihoods as equipment and tools used in tilling the land for rice, coconut, vegetables and fruit, were lost. Many also lost their micro-enterprises that included selling food and handicrafts at bazaars. On the domestic front young women survivors bear the burden of looking after children left motherless, cooking, keeping house for a few households and in some cases entering marriages, regardless of whether they consented. In some instances there have been reports of younger women being harrassed into marrying widowers and asked to bear children with shorter intervals to make up for the losses of children due to the tsunami. There are also instances of domestic violence and incidents of forcing widows into prostitution.[6]

A coordinator of Aceh's Trauma Recovery and Psychosocial Intervention Foundation cited the example of a village of 15 women survivors who were doing the work for the other 185 who died, to explain why women's workload had increased substantively, since the tsunami. She said women survivors were very busy — washing, cooking, serving the children and the men and the women had little time to think of their own pent-up emotions. One concern that activists have raised is the mental health

Photos courtesy of Mercy Relief

Women in shelters.

of women survivors to which the Indonesian Ministry of Health confirmed that many psychologists had perished in the tsunami and that 320 community leaders had been trained to provide psycho-social counselling.[7] It is estimated that about 400,000 Acehnese could be suffering from trauma-related stress disorders and that many of them are women.[8] Many have been encouraged to seek solace in religion.[9]

Immediately after the disaster, women also suffered from the preaching of some conservative *ulama* (religious leaders) who advocated that the tsunami was a result of women's sins. These *ulama* believed that women in Aceh had not followed the strict Islamic law and as a result more

Women managing the distribution of aid in local communities.

Photo courtesy of Mercy Relief

women had died in the disaster. In some cases there were reports of women being paraded in the village for not wearing headscarves in the hope that they would change their ways.[10] To respond to this preaching, both local and international NGOs have been countering to help women cope with their sense of shame and guilt.[11] Yet, even a year after the disaster some conservative *ulama* in some *kampong* (villages) still insist that women ought to be blamed for the disaster. The local and international NGOs need to be vigilant and constantly make an effort to work with the community to counter such a claim.

Women Helping Women

Women have been struggling to cope with the disaster as well as the new responsibilities. Women have expressed through local NGOs that they felt excluded from the decision-making processes on reconstruction efforts.

Women were upset over the barracks that had been built as they were too small and lacked privacy: the plywood walls were too thin and Peeping Toms had already drilled holes into them. They were also upset at the toilets that had been built outside the barracks and the distance between the male and female toilets were too close. In some cases the walls of the wash areas did not reach up to the roof, leaving gaps at the top that allowed for women to be peeped at as they showered.

"The toilet is too far from our rooms, so if you have to go outside at night to use the toilet, it is too dark to feel safe and there is no privacy in the houses," said Ms Hadijah, in an interview with a reporter six months after the tsunami.[12]

Most local NGOs in Aceh were women-based organizations. It was therefore easy for Flower of Aceh to convene a coalition of six women's groups in early 2005 to look at how women could be helped.[13] The result was the setting up of a Crisis Centre to help women cope with the trauma of loss and to refer women to camps and field hospitals. The coalition also organized a meeting to be part of the decision-making processes on recovery and reconstruction efforts. More than 400 women from 21 districts in Aceh attended the meeting[14] and the recommendations were submitted to the Aceh Reconstruction Agency. Top of that list was the setting up of the *Balai Inong* or Women's Houses, for women to meet, network and work together on projects.

The Executive Director of Unifem, Ms Noeleen Heyzer, who was making field visits to all affected areas also prompted major meetings with women's groups at the grassroots level. She advocated for women to be part of the solution process — to re-establish livelihoods, to work towards appropriate and adequate housing, to establish rights to land titles and to interact with local and national bodies in reconstruction efforts.[15]

These local NGO and international NGO recommendations were accepted by Dr Kuntoro Mangkusubroto, Director of the Badan Rehabilitasi dan

Photo courtesy of Lee Poh Onn

An ordinary *warung* (stall), typical of a small business often run by women.

Rekonstruksi of Aceh and Nias (BRR), an organization that was set up by the Indonesian Government on 30 April 2005. Dr Kuntoro also appointed Unifem as advisor to liaise with the Aceh Women's Council, a network created by the congress to represent the women of Aceh.[16]

There were also many efforts by women's groups and individual women to help the female population in the tsunami affected areas. In March 2006 on International Women's Day more than 1,000 Acehnese women walked through the streets of Banda Aceh to raise awareness to issues of domestic violence.'[17] Women's groups here and in other countries collected sanitary napkins[18] and headscarves in response to the appeal from women's groups in Aceh. Individual stories of women devoting their resources, time and funds to help other women are still apparent. E-mails played a big role in raising funds which were sent, for example, from one concerned individual that then made its rounds with money being sent in trust to a woman in Jakarta who then made the trip to Banda Aceh to give the money to a women's NGO.[19]

At the state level, the BRR is stepping up efforts to set up micro finance and micro enterprises to help women and men regain livelihoods as part of the rebuilding process as more than 100,000 small and medium enterprises had been lost in the tsunami.

There are also local efforts to encourage businesses — on International Women's Day 2006, the International Labour Organization (ILO), Indonesian Employers' Association, BRR and Unifem celebrated by promoting role models for women entrepreneurs. Around 45 women entrepreneurs in Aceh showcased their work in food production, catering, transport, restaurants, beauty salons.[20]

There are also Singapore-based efforts to help women rebuild their lives. Three Singapore hospitals under a $2 million training programme under an Aceh-Singapore Technical Assistance Programme, currently provide training for 200 doctors and nurses. A training centre will also be set up in Banda Aceh for more nurses, primarily women to pick up such skills over the next three years.[21] A maternity clinic to help women is being built by the Singapore Red Cross to help women in Meulaboh.

Women and children collecting containers of water at temporary housing.

Photo courtesy of Lee Poh Onn

Most efforts are to provide adequate shelter to the displaced people.

The Challenges

Despite all the progress made in the reconstruction and rehabilitation, many challenges remain. An important challenge is to keep corruption at bay and create systems for seamless coordination among the various groups, including both local and international NGOs, in reconstruction and rehabilitation.

Some groups want to pick and choose the villages they prefer to work

Looking forward to a brighter future.

Photo courtesy of Lee Poh Onn

in, causing saturation in certain areas and leaving others with no service at all.[22] There is competition among villages, too, for work with particular NGOs and INGOs. Many Achenese, having grown used to the monetary handouts, have grown dependent on them. Critics condemn this aspect as being alien to Acehnese culture and are upset that they are not being encouraged to become self-sufficient.[23]

There are 124 international NGOs (INGOs) and 430 local NGOs working in Aceh. Overnight INGOs offices grew to staff strengths of 400 to 500.[24] Yet in spite of this volume of expertise, according to the BRR there are still 67,500 people living in one-year-old tents and another 50,000 in government-built temporary barracks. People are losing faith at the slow pace of reconstruction efforts and in their villages. Village leaders who are the intermediaries between the government, NGOs and INGOs are facing a hard time in negotiating with the people to be patient. This is because too many INGOs and NGOs have made promises together with village leaders, making it difficult for the BRR to enforce a systematic approach. There are also mismatches in helping the people with some receiving too much, inappropriate aid or too little.[25] This has caused dissatisfaction among the people. There are also concerns about how people are going to be helped out of poverty with the costs of living escalating as people try to cash in on the funds available. Today to rent a house in Banda Aceh is ten times more costly than before the tsunami.[26]

Another challenge is the issues with the military and past rebel members who still make demands on villagers by acting as middlemen and expect a commission or who merely extort money from the villagers.[27] There are also cases of villagers still being beaten up by soldiers or civilians. INGOs and NGOs, too, speak of the need to negotiate with the military.

Nevertheless, in spite of these challenges the women's groups continue to advocate for greater involvement in the reconstruction and peace building effort of Aceh. Acehnese culture has indeed strong female role models and leaders as part of the tradition. The scope to recover and enhance this role should continue.

Notes

1. Scott Baldaurf, *Christian Science Monitor*, 21 July 2005.
2. Flower Aceh, established in 1989, is the longest-running NGO working with women in Aceh. Based in Banda Aceh, Flower Aceh has a dual mission of building justice in social systems in Aceh and Indonesia and empowering women in the social, political and economic spheres.
3. Ms Suraiya Kamaruzzaman, founder and chairperson of the Flower Aceh movement, presented these statistics in a paper, "War and Disasters: Rebuilding Nations", at the Institute of Southeast Asian Studies, on 18 March 2005. These same figures have been cited in later newspaper articles and reports by various organizations.

4. Oxfam Briefing Notes, "The Tsunami's Impact on Women", March 2005.
5. Ibid.
6. Ilaria Sisto, "Tsunami One Year After: Socio-Economic Impacts in Banda Aceh" (FOA Gender and Population Division).
7. Richel Dursin, "Tsunami Impact: Mental Health of Aceh Women Survivors Overlooked", IPS, 6 April 2005.
8. Ibid.
9. Ibid.
10. Nick Meo, "Tsunami was God's Revenge for your Wicked Ways, Women Told", *The Times*, 22 December 2005.
11. Global Exchange Report, "Aceh: Peace After the Waters? Challenges of Reconstruction and Peace One Year Later", March 2006; <http://www.acheh-eye.org/data_files/english_format/peace_process/peace-process_data-eng.html>. Global Exchange, founded in 1988, is a membership-based international human rights organization dedicated to promoting social, economic and environmental justice around the world.
12. Scott Baldaurf, "It Takes A Woman to Rebuild a Village", *Christian Science Monitor*, 21 July 2005.
13. <http://www.womensedge.org> The Women's Edge Coalition was created in 1998 to address the critical economic role that women play in developing countries, and to recognize that their voices were not being heard by U.S. decision-makers creating policies and enacting legislation that affects women worldwide.
14. Unifem, "Aceh: Tsunami's Women Survivors Demand Greater Role in Recovery and Reconstruction Efforts", 23 June 2005.
15. Jodie McGrath, "Unifem Calls for Greater Role of Women in Recovery and Reconstruction Efforts", 28 June 2005.
16. Ibid.
17. <http://www.apwld.org> Asia Pacific Forum on Law and Development is an independent NGO that is committed to enabling women to use law as an instrument of social change. It has consultative status at the Economic and Social Council of the UN (ECOSOC).
18. The National Committee of Unifem, Singapore, organized a collection of sanitary napkins in Orchard Road for the women in Aceh.
19. An e-mail from Singaporean Nancy Chng that made its way to many friends around the world who sent money to Jakarta-based Ms Linley York and a nun who was working with the Femina Group. More than S$30,000 was raised through this e-mail making its rounds.
20. Presentation of Best Women Entrepreneurs of Aceh Awards, 2006, press release, ILO, 8 March 2006.
21. <http://www.redcross.org.sg/tsunamirelief_projects_m.htm>; <http://www.alexhosp.com.sg>.
22. Global Exchange, "Aceh: Peace After Waters? Aceh: Challenges and Reconstrution and Peace One Year Later", <http:/www.actionaid.org.uk>, March 2006.
23. Ibid.
24. Ibid.
25. Adam Burke and Afnan, "Aceh: Reconstruction in a Conflict Environment; Views from Civil Society, Donors and NGOs", DFID, 22 August 2005, pp. 14, 16, 20, 23, 31.
26. Ibid.
27. Ibid.

5

Meulaboh

Lee Poh Onn

Introduction

The role of Singapore agencies[1] in aiding victims of the tsunami in Aceh, in the relief, rehabilitation, and reconstruction process of the region involved participation by the people of Singapore at many levels: government to government, non-government to non-government, and government to non-government.[2] Ordinary Singaporeans were also extremely forthcoming in donations, monetary and in kind, towards victims affected by the tsunami not only in Indonesia, but also in other affected countries like Sri Lanka, Thailand, Malaysia, and the Maldives. The response of Singaporeans was unprecedented; possibly because events of such a gravity have rarely occurred around the region, and especially in neighbouring countries which have very close ties with Singapore.

Ordinary Singaporeans came out in droves to volunteer at the various relief centres in Singapore: to collect clothing and foodstuff, and to provide assistance and support to the various Singapore NGOs busy with the rescue and relief operations. For example, at the office of Mercy Relief, volunteers came to offer assistance in gathering foodstuff, helping in its day-to-day operations, and to assist

Facing page: The new Meulaboh Pier. *Photo courtesy of Lee Poh Onn.*

in whatever matters that they could. Mr Izuan Rais of Mercy Relief recalls that these volunteers displayed a compassion never seen before. Their willingness to come together, their solidarity, and their selflessness and willingness to go out of their "comfort zones" to do whatever they could to help people in the tsunami struck areas really left a deep impact on his mind. Over 2,000 Singaporeans from all walks of life came to pack and separate clothing items at the Singapore Land Authority-Mercy Relief Collection Centre at Lowland Road from January to March 2005.[3] Elsewhere, relief supplies were overflowing into the Indonesian and Sri Lankan embassies in Singapore, with many Singaporeans giving up their holidays to assist in packing such supplies, volunteering to go overseas, and also helping to collect monetary donations for the disaster victims.[4]

Not only were donations and offers for help pouring in from the public, but the Singapore Government (Singapore Armed Forces and Singapore Civil Defence Forces) also actively provided relief and logistical support for the

The Meulaboh Pier that was completed on 3 April 2006. The new pier gives residents a logistics lifeline.

Photo courtesy of Lee Poh Onn

Photo courtesy of Lee Poh Onn

The Intensive Care Unit of the Meulaboh General Hospital. Donors are the Government of Singapore, Temasek Holdings and Singapore Red Cross Tsunami Fund with the following as supporting organizations: Singapore Red Cross, Mercy Relief, Palang Merah Indonesia, Bupati Aceh Barat, and Dinas Kesehatan.

disaster struck areas in Aceh, particularly in Banda Aceh and Meulaboh in Indonesia, and Phuket in Thailand.[5] The Republic of Singapore Air Force (RSAF) aircraft also flew supplies and support missions to Sri Lanka, the Maldives and Mauritius.

Meulaboh, the capital of Aceh Barat (West Aceh), is situated along the west coast of Sumatra and was one of the hardest hit areas. However, the focus on Meulaboh does not imply that the efforts of Singapore Agencies and NGOs were negligible elsewhere. The area of Meulaboh was selected because it was the area where the Singapore Armed Forces (SAF) concentrated its rescue and relief efforts when the tsunami struck in December 2004, and as such provides a good basis for beginning a study to examine issues involved in coordinating development and rehabilitation and reconstruction in one area.[6]

The contributions and challenges faced by these agencies is examined using a broad development economics approach, where rehabilitation and reconstruction involves not only restoring the economic infrastructure but also a restoration of the socio-economic infrastructure (hospitals, roads, schools, mosques, orphanages) that was important to society and individuals in Meulaboh. To what extent did the various agencies restore the socio-economic infrastructure? What are some of the challenges faced by these agencies in the rescue, rehabilitation, and reconstruction process? Could some common lessons be drawn from their experiences?

Scale of the Disaster

When the tsunami struck Indonesia on 26 December 2004, the magnitude of the damage was unforeseen.[7] About 167,000 went missing or died, over 500,000 individuals were displaced from their homes, around 80,000 to 100,000 houses were fully or partially destroyed, 3,000 kilometres of roads became impassable, fourteen of the nineteen seaports were damaged, eight out of the ten airports were destroyed, more than 2,000 school buildings were damaged, approximately 2,500 teachers died, more than eight hospitals were damaged or destroyed, the productive sectors of the economy were estimated to be damaged costing US$1.2 billion, 4,717 coastal fishing boats were destroyed, 60,000 farmers were displaced when over 60,000 hectares of agricultural land was destroyed, and 100,000 owners of small businesses lost their livelihood throughout Aceh and Nias.[8]

Of the total population in Aceh Barat of 176,586 before the tsunami, 12,873 individuals lost their lives, 2,403 went missing, and 72,806 became homeless.[9] Over 10,738 houses were completely destroyed, 1,892 units heavily damaged, and 1,903 damaged.

The subdistrict of Johan Pahlawan where Meulaboh is situated suffered the heaviest damage in terms of housing. Over, 5,640 houses were destroyed.[10] As at 20 June 2005, over 9,603 individuals were still living in barracks, and 12,023 individuals were living in tents in Aceh Barat.[11] In Meulaboh, which had a population of around 40,000, it was estimated that a quarter of its population perished in the disaster.[12]

Rescue and Relief

When Singaporeans heard about the earthquake and tsunami which had hit several countries in the region, no one could envisage its magnitude or the number of lives that were lost in this calamity. By the evening of 26 December 2004, deaths were reported in Bangladesh, India, Indonesia, Malaysia, the Maldives, Myanmar, Sri Lanka and Thailand. Three countries (Indonesia, the Maldives, and Sri Lanka) declared a state of emergency in their territories where the tsunami had struck.[13]

When the disaster stuck initially, the world could not fully grasp its enormity or extent, until reports and information started trickling in days later. Most of the news was at best sketchy, at that time conflicting and confusing. Reports from the Indonesian authorities were flowing in but the information was not sufficient to know precisely where aid and rescue operations could be effectively deployed. Even then, the full extent of the disaster and its impact on the people of

Indonesia was not to be completely grasped until much later.

The Singapore Armed Forces (SAF) and Operation Flying Eagle (OFE)

On the night of 26 December 2004, Lieutenant-General Ng Yat Chung, the SAF Chief of Defence Force, instructed his staff to arrange for a Joint Services Conference (JSC) the next morning for senior officers in the Army, Navy, and Air Force to discuss how the SAF could be involved in the rescue operations.[14]

SAF has always had on standby Operation Flying Eagle (OFE) for carrying out humanitarian aid missions to disaster affected countries. This consisted of a RSAF C-130 (Hercules) transport aircraft and a 25-person medical team of doctors and combat medics together with stores, equipment, and light cross-country vehicles which could be sent for disaster relief operations anywhere in the world. In the present situation, it was soon realized that a single C-130 consisting of a 25-person medical team would not be adequate. This was clear even before the situation report was presented to the JSC on the morning of 27 December 2004.

After the JSC, the SAF decided that it would be more appropriate to deploy three components to any of the recently disaster struck areas because of the magnitude of the disaster. There was to be a C-130 to transport medical supplies, a second C-130 to transport a 25-person medical team, and a Landing Ship Tank (LST), with helicopters to assist in the search and rescue mission.[15] Activation came on 27 December when the RSS Endurance (LST) sailed to the Tuas Naval Base to be on stand-by for rescue and relief operations.

Launch of Operation Flying Eagle

Operation Flying Eagle (OFE) was activated on 27 December 2004, a day after the occurrence of the tsunami. On 28 December 2004, SAF units had official orders to move into high gear. However, there was still one missing component: information was incomplete and a fuller picture of the disaster in the various countries was yet to be available.

The SAF mission planners were still deciding exactly where OFE would be most needed.[16] This decision could only be made when firsthand information became available where an SAF officer was in the disaster area itself. At that point, information on the affected areas was still scarce although the SAF realized that the disaster was indeed catastrophic.[17]

In the meantime, a smaller deployment consisting of a C-130 was waiting for clearance from Indonesia. The first clearance came on the afternoon of 28 December 2004. The SAF immediately dispatched the first catchment of relief supplies through the RSAF C-130 which included tents, groundsheets, blankets as well as food and medicine. During this mission to Indonesia, Colonel Toh Boh Kwee, who was its commander, also visited the Command and Control Coordination Centre in Medan which had just been set up to spearhead relief work in the devastated areas.

Photo courtesy of Lee Poh Onn

"Together with Aceh, to the Future": a girls orphanage project managed by Mercy Relief at Simpang Temasek, Meulaboh.

On 29 December 2004, information was quickly incorporated into the rescue plans of the SAF, in line with its focus of ensuring that the humanitarian task force was to be sent to an area where assistance was most needed. On 30 December 2004, Prime Minister Lee Hsien Loong and Minister for Defence Teo Chee Hean, together with other ministers directly involved in Singapore's tsunami rescue operations, gathered for the government's first media conference on the disaster.[18]

After discussion with the heads of government of Indonesia, Malaysia, and Thailand, the Singapore Government decided to focus on Indonesia and

Thailand as the disaster was of greater severity in these countries. The Minister for Defence Teo Chee Hean then mentioned Meulaboh, a place that had been completely devastated and cut off from the rest of Sumatra except through air. It was after discussions with the Indonesian authorities (Tentara Nasional Indonesia or Indonesian National Military), that Singapore decided to focus its relief efforts on Meulaboh. The Singapore Government nevertheless continued to work with the Thai authorities in Phuket but on a smaller scale.

The Humanitarian Assistance Support Group (HASG) in Meulaboh, Aceh Barat

Meulaboh subsequently became the place of focus for the SAF and Singaporeans involved in the relief efforts. The OFE was going to be on a scale never before undertaken by the SAF. The officer given command of the Humanitarian Assistance Support Group (HASG) was Colonel Tan Chuan Jin, Commander of the 7 SIB, a Guards Unit. Although the SAF had a template of "standard ready to deploy package and teams", Colonel Tan Chuan Jin held to the view that it was important to be "flexible and adaptable to the situation" and to uphold the intent of focusing "on aspects of assistance where the SAF would be most needed".[19]

As such, it was important to ascertain what was needed by the Indonesians in Meulaboh and then to work towards supplying and meeting these needs; matching demand with supply. Colonel Tan was clear about the broad strategies, basic principles, and the overall objectives and outcome of the mission. The challenge was in effectively applying this to the rescue mission, and achieving the highest possible value-added outcome within the given parameters. Safety was also a concern to Colonel Tan. The rescue operation had to be carried out at a pace that was sustainable for the SAF personnel. Awareness of the religious and cultural sensitivities was another important factor.

A first priority was to break down the rescue operation into manageable concerns and to devise solutions that would solve each of these concerns effectively. Then the challenge was to combine these individual solutions back into the big picture to resolve the challenge in its entirety.

In this HASG exercise, Colonel Tan also mentioned that the SAF's mindset was that of a "friend working closely together to solve a problem". Touching base with key Indonesian and Singapore personnel in the rescue mission at Banda Aceh was vital at that point in time — to seek feedback and to be keenly aware of the situation and what was needed on the ground. Colonel Tan also communicated with the Indonesian NGOs to understand the situation and to work on what they also felt was needed. This "human-to-human" communication element proved to be one of the crucial elements to the success of the overall rescue mission.

Colonel Tan worked alongside his Indonesian counterpart Colonel Geerhan Lantara in this HASG.[20] Colonel Tan mentioned that the role and support provided by the Tentara Nasional Indonesia (TNI) in working closely with the SAF personnel was very important. Working with the TNI also meant attending their meetings and respecting the important feedback that the Indonesian army personnel had to offer. In this respect, the goodwill and regular training and exchange exercises held between the military of Indonesia and Singapore over the years played a very important role in ensuring that both parties could work well together in such a massive and unprecedented operation.[21]

Colonel Tan's first priority was to create access to the area from sea and air (opening up a much larger helicopter landing site). He was in the advance party to ascertain what was happening on the ground before the Landing Ship Tank (LST) RSS Endurance arrived at Meulaboh's shores. The RSS Endurance left Singapore for Meulaboh on 31 December 2004 and arrived on 2 January 2005 carrying 470 people including a 33-strong medical team, Singapore Civil Defence Forces (SCDF) members, combat engineers, divers, and guardsmen. In terms of machinery and supplies, it carried 51 vehicles and heavy equipment like bulldozers, forklifts, excavators, dump trucks, mechanical shovels, generators, and 350 pallets and crates of relief supplies.[22]

Arrival of the Landing Ship Tanks (LSTs): RSS Endurance, Persistence and Endeavour

On 3 January 2005, landing sites had to be set up first before supplies and relief operations could commence. The bulldozers then started with the clearing work. A field hospital was also set up in Meulaboh at the displaced peoples' camp. At the Meulaboh General Hospital, a surgical team started operations at its medical centre. The doctors and medics from the Singapore Armed Forces (SAF) together with the Singapore Civil Defence Forces (Disaster Assistance and Rescue Teams) (SCDF) and volunteer doctors and nurses were the first foreign team to provide deployments and also to set up medical operations at Meulaboh in Aceh.[23] On 6 January 2005, the RSS Persistence arrived at Meulaboh carrying 196 men, heavy equipment and four containers of supplies including 200 cement packs, and 27 pallets of rice, water, milk and Milo. The portable water-purification plant producing 4,000 litres of clean drinking water also started operating on the same day. The SAF specialists also helped in restoring the mobile phone services in the area.[24]

The Singapore ships and helicopters made it possible to establish a field hospital and water purification plant just two weeks after the occurrence of the calamity. The SAF Chinook also assisted in setting up the Telkomsel GSM base station into Meulaboh.[25] The Chinooks are able to take four times more load than medium-lift helicopters and they are also able to fly further. They were literally the only medium and heavy lift helicopters in Aceh until 15 January when the U.S. Marines started flying in their CH-46s. The Chinooks also lifted the Indonesian Army's satellite communication systems, a huge parabola dish, and a team of people from Banda Aceh to Meulaboh.[26]

The RSS Endeavour, the third LST, arrived in Meulaboh on 16 January carrying NGOs and relief supplies. NGOs included those from the Singapore Red Cross (SRC), YMCA of Singapore, Touch Community Services International, Mercy Relief (MR), and the National Volunteer Philanthropy Centre.[27] The SRC was in charge of coordinating volunteers on board. The SRC worked closely and in consonance with the Indonesian Red Cross before and during this mission. There were initially some challenges faced by the SRC in operating in Aceh especially when around 35 per cent of the top tier members of the Indonesian Red Cross had perished in the disaster. Nevertheless, these hurdles were eventually overcome when operations proceeded on the ground. Like the SAF, the SRC was equally overwhelmed by the scale and magnitude of the devastation when the news hit Singapore's shores in December 2004.

Eventually, however, with more information coming in, the SRC managed to identify needs of the disaster victims and how Singaporeans could contribute to the relief and rescue efforts. In terms of its rescue and relief operations, the SRC followed a structured and measured approach that did not jeopardize the lives of its volunteers. Security was carefully considered before the SRC sent its volunteers to assist in OFE in January 2005 alongside the SAF's RSS Endurance.[28] Religious considerations and cultural sensitivities were not overlooked when undertaking the rescue operations; and the SRC also felt that it was important to instil into its volunteers the awareness that the local people were still grieving and in shock. With the benefit of hindsight, Mr Lim Theam Poh now believes that this is an important aspect that must always be emphasized to all workers in future disaster rescue operations.[29] The SRC also sent medical teams to work with the Indonesian Red Cross up to September 2005.[30]

Mercy Relief staff were among the NGOs in the RSS Endeavour. As Mr Izuan M. Rais pointed out, it was important to instil into the minds of the volunteers beforehand that they had to be "sensitive to the local context and have an open mind about the way things get done in a particular area.[31] Not everything can run like clockwork especially in disaster situations." Meeting the food, water, and shelter needs of the people was of top priority. Working closely with the local government was another important factor as well as mobilizing and matching needs. It was also crucial to send someone to the ground to gauge the situation and the needs of the people before proceeding further with relief operations and to have an exit plan.

During the OFE, three out of the four Singapore's LSTs were in Indonesia at one time. On 19 January 2005, combat engineers completed the construction of the modular building which was used to coordinate relief operations in Meulaboh. The HASG departed for Singapore on 21 January 2005. After completing the rescue mission, Colonel Tan noted that it was important to leave in place a working system that is sustainable and to provide also some semblance of order to the town so that recovery could continue to proceed.[32]

Reflecting on the achievements of the mission and lessons learned in 2004, Colonel Tan noted that the SAF personnel were very competent and focussed in terms of the execution and pace of the mission; the rigorous training over time had paid off with good results. It also showed that the resources and training to enhance SAF's operational readiness had been effective and was evident in the rescue and relief operations in Aceh.[33]

Medan, Banda Aceh, and Meulaboh

Although Singapore's rescue efforts were focussed on Meulaboh, Banda Aceh was not overlooked. The SAF medical team led by LTC (Dr) Fan Swee Weng arrived in Banda Aceh to offer medical services to the people there.[34] In close coordination with the Indonesians, the medical team set up a clinic at Ulee Kareng and began work effectively in the town.[35] All in all, six Chinooks, two Super Pumas, three helicopter landing ships (LSTs) and one mobile air traffic control tower together with more than 900 personnel were deployed by the SAF to Banda Aceh and Meulaboh in December 2004.[36]

Of noteworthy mention are the efforts of the SCDF efforts in Banda Aceh led by LTCs Yazid Abdullah and Yap Kok Boon.[37] The first contingent was led by LTC Yap from 29 December 2004 to 7 January 2005, the second contingent by LTC Yazid from 7 to 11 January 2005 in Banda Aceh. LTC Yazid recounts the shock he felt at the degree of destruction in Banda Aceh when he first arrived there. Both LTCs were there to help survivors and to locate bodies trapped under the rubble. The team led by LTC Yazid was also there to help survivors and facilitate the treatment of refugees at the Secata TNI camp.

By the end of the three-week operation in January 2005,[38] there were 1,700-strong SAF contingents in Medan, Banda Aceh, and Meulaboh. In Meulaboh, land and air bridges were created by building two beach landing points and seven helicopter landing sites. The SAF field hospitals in Meulaboh and Banda Aceh treated 5,174 patients. The LSTs ferried 190,000 kgs of relief supplies. The RSAF Chinook and Super Puma helicopters flew 143 missions, ferried 150,000 kilograms of relief supplies and carried 2,500 passengers. The C-130 and Fokker-50 aircraft flew 76 missions, ferried 240,000 kilograms of relief loads and carried 1,200 passengers. In terms of infrastructure, engineers built a modular building and two logistics storage areas. Such contributions were indeed modest compared to say what the United States contributed, but the early timing and precision of planning made a

Town activities in Meulaboh.

Photo courtesy of Lee Poh Onn

big difference to the rescue efforts as the people of Meulaboh were running out of water, food and medicine, providing niche capabilities that were vital to the rescue and relief operations. The SAF also ensured that its relief supplies were sent to individuals that needed it most. Relief efforts were also applied in a timely and effective manner. The SAF helped in areas where expertise was present, recognizing that it had to maximize its capabilities in areas where it had a natural advantage.

Singapore Agencies

Singapore's civilian entities were also involved in the rescue efforts. Singapore Airlines offered carriage of relief supplies to all affected countries, the Civil and Aviation Authority handled ad-hoc non-scheduled cargo flights carrying relief aid at Changi Airport, waiving numerous charges. The Port of Singapore Authority waived handling charges for ships carrying relief aid.[39] The Singapore Public Utilities Board (PUB) sent bags of water to Aceh upon news of the disaster.

In addition, it also worked with local private water firms to set up portable desalination plants to provide a sustainable supply of clean water to the populace affected by the disaster and also during the reconstruction process. The PUB sent a portable plant with a capacity of 50 cubic metres per day to Meulaboh. Keppel Engineering fabricated a plant with a capacity of 600 cubic metres per day for Aceh shipped out in mid-January 2005.[40]

The Singapore Red Cross, Singapore International Foundation, YMCA, Mercy Relief, Care for the Elderly Foundation, City Harvest, Touch Community Services, Singapore Buddhist Federation, Ren Ci Hospital and Medicare Centre were some of the NGOs that sent relief missions with teams of nurses and doctors to Indonesia.[41]

Besides sending relief missions, many organizations, including those mentioned above, were involved in raising funds in Singapore for the tsunami victims. These included the Singapore Red Cross, Salvation Army (Singapore), YMCA (Singapore), Medical Foundation, Touch Community Services, Mercy Relief, Bright Hill Monastery, City Harvest Community Services, World Vision (Singapore), Habitat for Humanity (Singapore), and the Singapore Scout Association to name a few.[42]

Most of these non-governmental organizations are Singapore establishments, although some have an international link.[43] The Singapore Red Cross, founded in 1973, specializes in relief operations during disasters and in health and welfare services. Mercy Relief was started in 2001 by the Adult Islamic Religious Students' Association of Singapore (Perdaus) to respond to natural disasters in Southeast Asia. The Medical Foundation was set up by the Raffles Medical Group in April 2003 to raise funds for needy patients. Habitat for Humanity is a Bangkok-based, non-profit Christian organization, set up in Singapore in January 2003. It builds houses for the poor. The YMCA is a Christian organization with worldwide affiliations, set up in Singapore in 1902. Touch Community Services was set up in 1999 as a non-profit voluntary organization conducting crisis relief in many parts of the world. The Singapore arm of the British Salvation Army was set up in 1935, and has conducted emergency relief programmes after typhoons and earthquakes in many parts of the world.

The Singapore International Foundation (SIF), a not-for-profit organization, was established in 1991 to build a corps of active global citizens for Singapore through active volunteering and networking. The Ren Ci Hospital is a non-profit voluntary organization founded in Singapore in 1994 to look after patients suffering from severe physical disabilities and terminal illness. The City Harvest Community Services is a non-profit social work agency founded in 1996. It runs multi-racial community services and rehabilitation programmes for the elderly, intellectually disabled, abused, terminally ill, youth at risk, and

Photo courtesy of Mercy Relief

The Indonesian Red Cross, or Palang Merah Indonesia, worked closely with the Singapore Red Cross.

Photo courtesy of Mercy Relief

The Singapore Red Cross delivering a truck-load of supplies.

prison inmates. The Buddhist Fellowship is non-sectarian and founded in 1988. In the past, it has undertaken relief work in poor communities. The Singapore Scout Association is a non-profit association set up in 1910 to create opportunities for youth to be active in society through community based projects. World Vision is a Christian relief and development organization which focuses on the well-being of children.

NGOs in Singapore went all out to assist the tsunami disaster victims, often in collaboration with one another. For example, when the disaster first struck, Mercy Relief together with the Singapore Soka Association and Young Sikh Association set up five collection centres to receive donations in kind.[44] Mercy Relief also mobilized over 100 volunteers to sort and pack donations in kind that were rejected by other relief groups due to the high costs of shipping and the complex logistics involved. Mercy Relief collaborated with the Singapore Petroleum Association and MediaCorp Radio to supply drinking water to the tsunami affected areas. Mercy Relief was the first NGO from Singapore to undertake reconstruction efforts in Meulaboh.

Agencies like Mercy Relief and the Singapore Red Cross, through its volunteers, helped to collect donated items including food, water, tents, generators and vehicles which were transported through Singapore's helicopter landing ships provided by the Singapore Armed Forces. The Singapore Red Cross also coordinated relief expeditions bringing volunteers from the various aid and charity organizations to Meulaboh, and ferrying medical supplies, water, batteries, face masks, blankets, and field tents.

Rehabilitation and Reconstruction

ASEAN Summit

Prime Minister Lee Hsien Loong called for an urgent meeting under ASEAN auspices on 30 December 2004[45] to hold a special summit under the United Nations to establish a special fund to help post-disaster relief and reconstruction efforts.[46] The Summit was subsequently held in Jakarta on 6 January 2005, hosted by Indonesian President Susilo Bambang Yudhoyono. Leaders of 26 nations and international organizations came together for the meeting, with US$4 billion pledged by governments and private donors.[47] In this summit, Singapore pledged US$10 million in aid of all countries affected by the tsunami.[48]

Following the summit, the Indonesian Government, together with the international community, prepared a damage and loss assessment on 19 January 2005. On the same day, Vice-President Jusuf Kalla stated in an interview published in the *Straits Times*[49] that the emergency relief, rehabilitation, and reconstruction phases were to run concurrently. The emergency relief phase which began at the onset of the tsunami was expected to last between one to three years. The next phase was rehabilitation which would last for one and half years. The third reconstruction phase would begin in the middle of 2005. As such, Indonesia is now at the rehabilitation and reconstruction phases with reconstruction expected to last for about five years. Around US$2 to 3 billion is needed for infrastructure and housing, electricity and power, seaports, bridges, hospitals and schools. Singapore was acknowledged for initiating the ASEAN Summit and for the SAF assistance in Aceh. Indonesia will be looking to Singapore for its expertise in town planning, logistics aid, air transport, and medical support.

In March 2005, the Indonesian Government under its National Development Planning Agency, BAPPENAS, prepared the Master Plan for Rehabilitation and Reconstruction of Aceh and Nias which was released on 26 March 2005.[50]

Badan Rehabilitasi dan Rekonstruksi (BRR)

On 30 April 2005, the Badan Rehabilitasi dan Rekonstruksi (BRR) or Rehabilitation and Reconstruction Agency in Indonesia was established to facilitate the rescue, rehabilitation, and reconstruction process of the tsunami affected areas. The BRR comprises three bodies: the Executing Agency (Bapel), headed by Kuntoro Mangkusubroto; a high level Advisory Board to guide the reconstruction strategy; and an Oversight Board to monitor activities, handle public complaints and to conduct audits. All three bodies report directly to the President of Indonesia. The BRR coordinates rather than implements projects, and receives and evaluates project proposals fundable by either government resources or from funds received from the international community.[51]

To date, the Singapore Government has given over S$70 million including an initial pledge of S$5 million for immediate relief, S$16.5 million for reconstruction, and S$44 million for deploying assets and personnel to all affected countries.[52] This sum does not include donations coming from the non-government sector. In total, over S$150 million has been collected for the tsunami disaster, making it by far Singapore's largest humanitarian endeavour.

The Tidal Waves Asia Fund (TWAF)

The Singapore Red Cross (SRC) relief fund tops at over S$88 million, raised through its "Tidal Waves Asia Fund" (TWAF).[53] More than half the funds raised are from individuals. Over half the funds have been used for reconstruction projects in regions affected by the tsunami.[54] About 65 per cent of the funds used for reconstruction has gone into projects in Aceh, the rest to the other tsunami hit areas. The Tsunami Reconstruction Facilitation Committee (TRFC) was set up in January 2005 by the SRC with various NGOs, VWOs and the Ministry of Community, Youth and Sports (MCYS) to manage the TWAF. LG (Retd) Winston Choo (Chairman, SRC), Associate Professor Lim Meng Kin (Council Member, SRC), Mr Chew

Hock Yong (Deputy Secretary, MCYS), Mr Liak Teng Lit (CEO, Alexandra Hospital), Mrs Tan Chee Koon (CEO, National Volunteer and Philanthropy Centre), Mr Sim Cheng Huat (Senior Consultant of International Enterprise (IE), Enterprise Advisory Department), Dr Tan Tay Keong (Executive Director, SIF), Mr Zulkifli Baharudin (Chairman, Mercy Relief) and Mr Eric Teng (President, YMCA Singapore) are members of the TRFC.[55]

How are projects vetted? Project proposals are submitted to the TRFC Secretariat who will then raise them to the TRFC members. The TRFC reviews project proposals to ascertain if projects effectively and efficiently meet the needs of society[56] in the tsunami struck areas before such projects can be approved for funding.[57] Projects that equip locals with knowledge and skills in re-establishing their lives and livelihood are of utmost importance also. Only project proposals for Sri Lanka, Indonesia, and the Maldives are considered. Joint projects between Singapore NGOs/VWOs with responsible agencies in the affected countries are favourably considered.

Once a project has been approved, the TRFC Secretariat of the SRC tracks and oversees the project until its completion. Upon completion, the NGO or VWO will be required to submit an audited statement of accounts and project closeout report. The whole process is undertaken in a systematic, transparent, and accountable manner. Mr Lim Theam Poh, Manager of International Services of the SRC, emphasized that "neutrality and impartiality" is an important element in considering the dispensation of funds to the various NGOs and VWOs.[58] Transparency in the process of allocating funds is another important criteria; donors can view the list of approved projects at the website of the Singapore Red Cross. Colonel Christopher Chua (Retd) also mentioned that the donations collected were used solely for TRFC projects, all interest earned from these funds were ploughed back into the TWAF, none of this funding is used for the day-to-day operational expenses of the SRC.[59]

The SRC is also involved in the long-term capacity building of medical staff in Aceh, and in developing the management skills of Acehnese to dock ships in the new Meulaboh pier through collaboration with personnel in the Port of Singapore Authority (PSA). About S$20 million out of the S$88 million of the TWAF had been allocated to NGO/VWO projects in Meulaboh as at September 2006.[60]

The Role of Singapore Agencies in Aceh

In January 2005, Temasek Holdings, also announced that it had set aside US$10 million for the immediate relief, short-term disaster recovery, and medium-term reconstruction support for Aceh and other tsunami hit areas.[61] For the immediate term, Temasek would work with the Singapore Red Cross, Mercy Relief and NTUC Fairprice to support their efforts to send food, water and medical supplies to victims in the affected countries in the region.[62] For the shorter term, support from Temasek includes the provision of community-support systems such as power and water supplies, doctors, and communications systems. In this respect, Temasek would work with Hyflux Ltd, which specializes in water-related technologies, to supply water-generating as well as water purification systems to Aceh. In the medium-term, Temasek Holdings will provide support in the reconstruction and rebuilding of basic infrastructure projects such as hospitals, schools, water, and power generators.

The Singapore International Foundation also joined in by raising funds through its Tsunami-Affected Communities Fund. The SIF's executive director, Dr Tan Tay Keong, told *Business Times* in January 2005 that the SIF planned to send 60 teams of about 15 volunteers each to areas ravaged by the tsunami to help with reconstruction efforts in April.[63] The SIF's Singapore Volunteers Overseas (SVOs), made up largely of specialists and professionals, also helped to build a potable water sanitation and distribution system for the Desa Sirombu village in Nias.[64] In the Aceh Province, SIF volunteers helped to construct a Community Centre-cum-Kindergarten and radio station. Refurbished and new mobility aids (including wheelchairs, commodes, crutches and walking frames) were also sent to medical institutions in the affected areas. Using the "training-of-trainers" approach, SIF volunteers taught Aceh NGO volunteers to maintain and refurbish wheelchairs in August 2005. The Move for Mobility Project is a collaborative initiative by the Society for the Physically Disabled, Singapore International Foundation, Mobility Aid Recycling Services, and Keppel Facilities Management and Operations Pte Ltd.[65] At Zainoel Abidin Hospital, a state hospital severely affected by the tsunami where about 50 per cent of the manpower perished, SIF volunteers trained new staff, and "provided the logistical knowledge to support operations to complement medical services". The project was completed in March 2006.[66] The SIF together with an Acehnese NGO, the Patimadora Foundation, also carried out a livelihood project to train Acehnese in tailoring skills in order to generate income for these families. These families are currently staying at the Internally Displaced Persons (IDP) camps. This project ended in March 2006.[67]

Mercy Relief, with funding provided by its corporate donors,[68] also entered into a contract with one of Indonesia's most established shipbuilders, P.T. Pal Indonesia, to construct and supply 10 fishing vessels to the communities in Aceh and Nias who have lost their livelihood due to the tsunami.

Medical teams from Singapore's KK Women's and Children's Hospital and the Alexandra Hospital stayed in the hospitals in Aceh to help staff cope with the large number of patients and also to give them time to find replacements for the staff lost to the tsunami. In the present rehabilitation and reconstruction efforts,[69] doctors and nurses from Aceh have also been funded by the TWAF to be trained in Singapore on the various aspects of health care, hospital operation and administration. The fund known as the "Aceh-Singapore Technical Assistance" totals about S$1.996 million. This month-long programme is in collaboration between hospitals in Singapore (Alexandra Hospital, KK Women and Children's Hospital and Singapore General Hospital) and hospitals in Banda Aceh and Meulaboh. Staff from Banda Aceh and Meulaboh receive training in Singapore in clinical, nursing, and management skills for health-care workers, and also know-how on rebuilding the training centres at hospitals in Aceh.

The Role of Singapore Agencies in Meulaboh, Aceh Barat[70]

On its own, the Singapore Red Cross is involved in the construction of a kindergarten and one maternity clinic in Meulaboh. The projects were still on-going as at end September 2006.[71]

The Singapore Red Cross together with Temasek Holdings, and the Singapore Government also assisted in

A commemorative plaque at the Meulaboh Pier that was reconstructed after it had been totally destroyed by the violence of the tsunami.

Photo courtesy of Lee Poh Onn

the recovery efforts by funding the construction of a new S$5.56 million pier at Meulaboh which was completed on 3 April 2006.[72] The 175 metre long reinforced concrete pier, able to handle 4,000 deadweight tonnes, will be the key entry point for essential supplies to the town of Meulaboh and the region.[73] The pier replaced the Dermaga ferry jetty which was damaged when the tsunami struck. The pier has the immediate task of facilitating the entry of reconstruction materials for Meulaboh and will thus speed up the rehabilitation of the devastated capital city. It will also aid in the construction of the trunk road from Banda Aceh to Meulaboh.

The Singapore Red Cross, together with Temasek Holdings and Mercy Relief are also funding a S$12 million project to modernize Meulaboh's General Hospital and provide healthcare training for Indonesian personnel.[74] This renovation and modernization is expected to be completed by the end of 2007.[75] The General Hospital has served the populace in Meulaboh treating about 200 patients per day on a staff of 12 doctors. With the reconstruction and redevelopment of the hospital which includes new facilities and equipment (increasing the number of beds from 80 to 144, and establishing clinical research labs and a new radiology department), the hospital will closely resemble private hospitals in Medan and Jakarta. Staff from the Alexandra Hospital, KK Women's and Children's Hospital and Singapore General Hospital have been providing training to the health-care workers in the Meulaboh General Hospital.[76]

Dr Haris M. Saputra, Director of the Hospital, spoke about the present infrastructure in the hospital and about future plans to upgrade such facilities by computerizing its day-to-day operations when the reconstruction of the hospital and its new wings have been completed. He noted that the refurbishment of the hospital will expand the hospital to over 140 beds with new departments in neurology and urology.[77] The current facilities and infrastructure at the hospital have been insufficient to meet the demands of the day.

An on-going staff restructuring programme is currently taking place and is expected to be completed when the new hospital is ready. The system of patient registration will be computerized, with staff receiving training on computer systems since July 2006. Not only will patient records be computerized, it is expected that the finance department and fee collection will also be computerized. Mr Rudi Ariyani, Administrator of Logistics and Training[78] of the hospital, was determined that a whole new mindset must be established for the hospital staff at Meulaboh — that the level of professionalism and dedication to patients must approach the benchmarks set by its Singapore counterparts.

Dr Haris also reflected on the tsunami and recalls the assistance (medical supplies and doctors and nurses) provided by the Singapore Navy, the SAF, and the SCDF on the fifth day after the tsunami disaster struck Meulaboh in December 2004. Since then, the Meulaboh General Hospital has continued to receive significant support from the exchange and training programmes between the Meulaboh General Hospital and the Alexandra Hospital in Singapore.

All in all, Dr Haris has a three-pronged approach for the future development of the hospital. The first consists of staff training and retraining of nurses and doctors through the exchange programmes in Singapore, Yogjakarta, Bandung, and Banda Aceh. The second is in upgrading the present equipment as well as the level of staffing of doctors and nurses. And the third consists of restructuring the infrastructure of the hospital and computerization of its services, records, and financial system.

In Meulaboh, Mercy Relief has also helped by constructing boarding houses, orphanages, and a school for children who have been displaced by the tsunami.

General Hospital in Meulaboh.

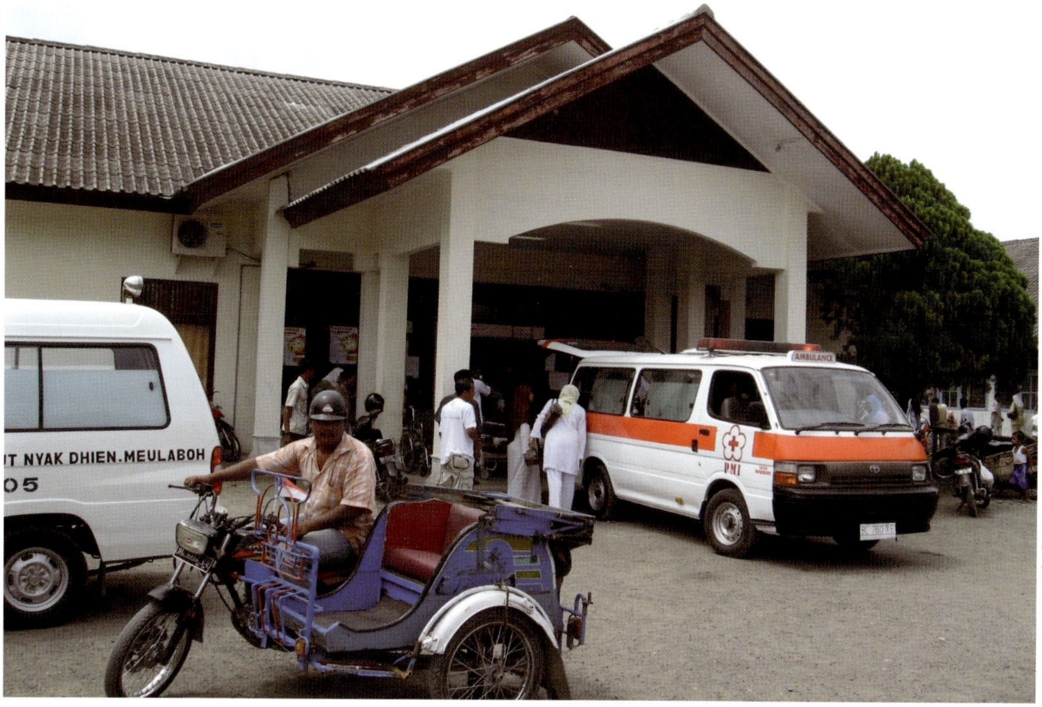

Photo courtesy of Lee Poh Onn

Photo courtesy of Lee Poh Onn

In September 2005, Mercy Relief signed a MOU for two new projects. One was in the reconstruction of the Babussalam Boarding School in the town of Meulaboh, and other was the Samatiga Boarding School, located 30 minutes from Meulaboh.[79]

Although one-off funding from the Jurong Country Club had significantly facilitated the construction of the new Babussalam boarding school, funding of about S$700,000 was still needed to complete its construction.[80] Half of the 284 boarders (made up of 127 girls and

Surau located at the Boy's Orphanage at Simpang Temasek in Meulaboh, a project managed by Mercy Relief.

Photo courtesy of Lee Poh Onn

157 boys) are orphans, and represent the poorest of the poor. Mr Edward Langton (Project Manager Meulaboh) and Mr Abdul Wahab (Liaison Officer Meulaboh) of Mercy Relief have been managing the project which began in November 2005. Mr Langton, through TRFC funding, has even completed setting up a small bricks factory at the vicinity of the boys section of the Babussaalam Orphanage to manufacture bricks for Babussalam's reconstruction, as well as for sale.[81]

For the Samatiga Boarding school, a new site has been secured for its construction which began in October 2005. There will be education and accommodation facilities which will include a school block, kitchen/canteen, dormitories, prayer hall, and a kindergarten. The project is funded by Yum! Restaurants International and the Chicago School of Business Alumni.[82]

Mercy Relief[83] has also successfully completed the restoration of the Muhammadiyah Orphanages, a multipurpose hall and a *surau* on 25 September 2005.[84] The multi-purpose hall was built for the needs of the Muhammadiyah but is also an income generator as it is sometimes rented out for functions. The girls orphanage (Panti Asuhan Muhammadiyah — Puteri) was supported by Commerzbank while the boys orphanage (Panti Asuhan Muhammadiyah — Putera) and the multi-purpose hall was funded through the TWAF of which the Singapore Red Cross is the custodian. The children living here are from all over Aceh. Altogether, there are around 24 female, and 26 male children.

Besides attending weekday schooling outside these orphanages, the children get to attend special classes conducted from time to time within the Muhammadiyah compounds consisting of computer skills, Koran reading, and learning the English language. Some also receive training in broadcasting (*ceramah*) as there is a radio station run by the Muhammadiyah (Radio Matahari) at its secretariat located in the vicinity of the boys orphanage. In addition to the orphanages, Mercy Relief with funding from the Lien Foundation has also completed the construction of the Sekolah Muhammadiyah.

Habitat for Humanity (Singapore) started a process of constructing low cost housing in Meulaboh in November 2005.[85] As an NGO, Habitat was registered in Singapore in 2003 by Mr Yong Teck Meng. The tsunami disaster was a catalyst for a steep learning curve which has proved to be invaluable for Habitat Singapore's rehabilitation and reconstruction portfolio in disaster areas. Habitat consulted closely with its counterparts Habitat Indonesia and Habitat Asia Pacific before proceeding with the construction phase.

In Meulaboh, Habitat is building 1,500 houses with funding coming from people in Singapore through the TWAF. Habitat received the largest disbursement of S$13.9 million from the Singapore Red Cross.[86] Mr Yong highlighted that it was important not to rush into a disaster area to build houses but to be thoroughly familiar with local conditions and the

Housing built by Habitat for Humanity.

Photo courtesy of Lee Poh Onn

cultural sensitivities first. Mr Yong also applauded the completion of the Meulaboh pier which would help to relieve the jam in supplying building materials and also lower transportation costs of these materials.

Ms Juliah Bee, Meulaboh-based Project Liaison Officer, agreed that materials for construction were much easier to obtain now that the Meulaboh pier had been completed.[87] Villagers in general have been very happy with the type of housing provided by Habitat for Humanity, and many have opted to be under Habitat's housing scheme rather than a scheme by other NGOs. Nevertheless, it was not under Habitat's purview to select who the beneficiaries are. Rather the decision to build houses in a particular location comes from the Bupati Aceh Barat which identifies areas where housing is needed and then allocates such areas to the different house-building NGOs.

Habitat is one of the top four house-builders in Aceh Barat. They are currently constructing houses in the Meurebo, Kuta Padang, Samatiga and Arongan Lambalek Sub-Districts. Habitat Singapore involves the owners and their family members to assist in the construction of their homes. Construction supervisors and managers ensure that overall quality standards are maintained. There is a process of consultation and discussion with the community before the houses are built. Habitat plans to complete two-bedroom units in these areas. On the matter of handling delays in supplies, Ms Juliah pointed out that the BRR has been extremely forthcoming in handling such matters and expediting any blocks in customs clearance and matters of the like.[88]

Mr Yong stressed that it was also important to educate donors about the realities on the ground, that the process of construction might not proceed smoothly due to many unforeseen circumstances. Donors' expectations have to be realistic, timings and delays can sometimes occur because of logistical difficulties and weather conditions. One major challenge was in estimating costs:[89] the price of housing materials has more than doubled because of inflation in Aceh. This has led to a downward adjustment of the

Photo courtesy of Lee Poh Onn

Fish market at Johan Pahlawan, Town Area, Meulaboh.

number of houses initially planned by Habitat from 3,000 to about 1,500 units in Meulaboh. Over 600 houses had been completed as at 28 August 2006 and the remaining due for completion soon after. A measure of Habitat's success has been the many instances where other NGOs have approached Habitat to assist them in their building projects.

Almost Two Years on…

The central market at Johan Pahlawan is again abuzz. The town is alive with people and activity. The Meulaboh pier is busy with shiploads of foodstuff and cement being unloaded. These signs bode well for Meulaboh but how will development proceed in the future? Will this opportune moment be fully seized by the people of Meulaboh? What are some of the remaining challenges?

Officers in the Bupati Aceh Barat were generally positive about the development that Meulaboh and Aceh Barat have achieved so far. The Bapak Bupati, Doctorandus Sofayanis, mentioned that the progress of rehabilitation and reconstruction in Aceh Barat has been "*cukup baik*" (good) so far.[90] In terms of the future progress of the area, Drs Sofayanis envisaged a multi-pronged development involving infrastructure (more roads and highways), education (expansion in the number of universities, secular and religious schools), communications and information and communications technology (increase in the use of the Internet and broadband technology). He also mentioned the support received from international organizations including those from Singapore that have played a significant role in facilitating progress in Meulaboh.

Drs Hasan Abdullah, Assistant to Economic Development and Reconstruction, Bupati Aceh Barat (Asisten Bidang Pembangunan dan Ekonomi Setdakab Aceh Barat) is involved in economic reform that is currently being pursued in Aceh Barat in the agricultural, fisheries, and small industry sectors.[91] In terms of agriculture, Drs Hasan mentioned that the Bupati Aceh Barat was targeting to develop the palm oil, rubber, buah pinang, rice, and cocoa sectors. Developing the fisheries sector was another priority. Drs Hasan pointed out that the Asian Development Bank was currently undertaking a survey on the fisheries sector.

Currently, the frequency of ships arriving at Meulaboh is still irregular. However, the completion of the Meulaboh pier was a significant move in the right direction for the development of the town. Unemployment levels and poverty rates are still high; nevertheless, such rates have been falling in recent months. The Bupati Aceh Barat has set a target of reducing poverty at one per cent per annum.

Mr Tueku Ahmad Dadek, District Officer of the Administration of the Bupati Aceh Barat, coordinates the activities of NGOs at the sub-district level of Johan Pahlawan in Meulaboh, where the agricultural and commercial infrastructure was destroyed by the tsunami.[92] Fisheries were damaged, traders lost their capital, the whole economic structure was, in effect, wiped out. The recent restoration of one out of the three ice factories in Meulaboh has been crucial for the redevelopment of the fisheries sector. Access to credit by traders have also been limited since the tsunami. However, some of the NGOs have assisted by providing micro-credit to these traders. Most significant is the scheme to allow drivers to own their *becak* (trishaws). This has increased the source of livelihood for 1,117 individuals. Previously, 1,017 traders had to rent their *becak*. Over 25 per cent of houses have been completed and about 12,000 individuals are still living in tents.

Haji Zaini Bashar, Head of the Muhammadiyah,[93] stressed that medical facilities need to be improved, that the educational standards could also be further raised, including the importance of using English in the community. Of special mention is the need for a public library in Meulaboh. Many books were destroyed during the tsunami and as such the children in Meulaboh are deprived of reading materials.

Near the ice factory that was restored at Johan Pahlawan, Meulaboh, and that has been crucial for the redevelopment of the fisheries sector.

Photo courtesy of Lee Poh Onn

Mr Syed Fadhlain, Muhammadiyah Youth Leader,[94] emphasized the importance of education for the youth of Meulaboh. He wanted the youth in Meulaboh to be a thinking generation, and also one that was fluent in English in order to facilitate communication with the international community. He felt that the current standard of English could be improved upon, and much remains to be done, as Aceh has been isolated from the world for thirty years. He also hoped that the youth exchange programmes between Singapore and Meulaboh will continue.

Photo courtesy of Lee Poh Onn

Interior of a mosque in Meulaboh.

Policy Recommendations

The Acehnese people are very determined to move on in the name of progress. They do not wish to be left behind. Acehnese are very eager to learn English, realizing that this is the language of international communication. English is taught in schools, but much can be done to improve its current level of usage as well as standards.

The provision of medical facilities and services can still be improved upon. In this respect, the exchange programmes between the hospitals in Singapore and the Meulaboh General Hospital is a move in the right direction.

Although progress in rehabilitation and reconstruction is evident in the capital of Meulaboh, further improvements in the infrastructure (roads, schools, communications, medical facilities), as well as speeding up the process of house building, and the provision of social services for the people in general and also orphans who are victims of the tsunami are needed.

The mention of Singapore to the people in Meulaboh is always accompanied by expressions of smiles, and gratitude for the assistance provided by the Singapore Government, Singapore agencies, and Singaporeans.[95] The continuing support from Singapore, financial and otherwise, will remain very important in the years to come.

Conclusion

The role of Singapore agencies has been crucial in the rescue, rehabilitation, and reconstruction process in Aceh and Meulaboh. The lessons learned have been numerous. The close relationship between Singapore and Indonesia, cooperation in military exercises, familiarity with culture and sensitivities, and human-to-human ties generated over the years, have enabled coordination and humanitarian operations to be applied in a timely and effective manner.

For the rescue and relief operations, a knowledge of the situation on the ground, military goodwill created over the years, having a drawer plan for humanitarian assistance (creating access, food, water, shelters, and medical treatment) adapted and flexible to the situation, touching base with leaders and people on the ground in Indonesia, working with the TNI, cultural and religious sensitivities, and the SAF preparedness and training programmes (operational readiness) over the years have been important factors crucial to the success of the mission.

For the rehabilitation and reconstruction process, knowledge on the ground, acceptance that things are done differently, familiarity with and understanding of the religion, culture and people, having clear project objectives, goals and an exit plan, are some of the important elements for success.

Notes

1. Information on the extent of the earthquake and tsunami drawn from the following reports: David Boey, *Reaching Out: Operation Flying Eagle SAF Humanitarian Assistance after the Tsunami* (Singapore: Ministry of Defence and SNP International, 2005); and BRR and

International Partners, *Aceh and Nias One Year after the Tsunami: The Recovery Effort and Way Forward* (Indonesia: BRR and International Partners, December 2005).

2. Singapore agencies in this chapter include the Singapore Government and its various agencies, Singapore voluntary welfare organizations (VWOs), and non-governmental organizations (NGOs) based in or with offices in Singapore.
3. Interview with Mr Izuan M. Rais, Head, International Relief Division, and Mr Mohd Tahar Jumaat, Disaster Response (Operation), International Relief Division, Mercy Relief, Singapore Office, on 20 September 2006. Mr Mohd Tahar Jumaat, a paramedic by training, was himself a volunteer during the rescue operations in January 2005 (RSS Endurance), and spent eight days there helping to clear the debris and assisting in providing medical aid to the people in Meulaboh.
4. "Indonesian Embassy, Sri Lankan High Commission Swamped with Donations", *ChannelNewsAsia.com*, 30 December 2004.
5. Government, VWOs, and NGOs active in the relief and rehabilitation efforts in Aceh through the Tidal Waves Asia Fund raised through the Singapore Red Cross include the Singapore Government, Temasek Holdings, Ministry of Foreign Affairs, Singapore, Singapore Red Cross, Mercy Relief, Singapore International Foundation, Marsiling YEC, Order of Malta, Raleigh Society, West Coast CCC, National Technological Singapore, NTU Muslim Society, Believe in Hope, Crisis Relief Society, Tomorrow's Hope, Catholic Archdiocese of Singapore/ Catholic Medical Guild of Singapore, Central Singapore CDC, Mobile Aid Repair Services (MARS) and Keppel FMO and the Society of Physically Disabled (SPD), Alexandra Hospital, KK Women and Children's Hospital, Singapore General Hospital, Water Initiative for Securing Health (WISH), World Toilet Organization, Evangelical Fellowship of Singapore, and Life Community Development Co Ltd. Please see List of Tsunami Approved Projects (Tidal Waves Asia Fund) at <http://www.redcross.org.sg/tsunamirelief_projects_m.htm> for more details. Information retrieved on 21 September 2006 by author.
6. A field trip was conducted in August 2006 in Meulaboh, Aceh Barat, to examine the extent of progress achieved in the area. Besides interviews conducted with Acehnese officials and Singapore NGOs in Meulaboh (Bupati Aceh Barat, Meulaboh General Hospital, Muhammadiyah, Mercy Relief, Habitat for Humanity, Singapore), interviews or discussions were conducted in September 2006 with individuals in Singapore agencies that had operated in Meulaboh (Singapore Armed Forces, Singapore Civil and Defence Forces, Singapore Red Cross, Mercy Relief, Habitat for Humanity Singapore, and Singapore International Foundation). The objective of these interviews was to gain insights into the experiences and challenges faced by these agencies in the rehabilitation and reconstruction process in Aceh. Thanks go to Mr Agung Wicaksono for assistance in translating and explaining Indonesian terms and phrases.
7. BRR and International Partners, op. cit., p. 16.
8. Ibid.
9. Ibid., p. 3.
10. Ibid., p. 6.

11. See Table 2 and 3 of ibid., pp. 4–5.
12. Boey, op. cit., p. 76.
13. Ibid., p. 21.
14. This section draws heavily from ibid., pp. 21–22. Thanks also to Colonel Benedict Lim, Director, Public Affairs, Ministry of Defence, for copies of David Boey, *Reaching Out: Operation Flying Eagle SAF Humanitarian Assistance after the Tsunami* (Singapore: Ministry of Defence and SNP International, 2005) which provides a detailed account of the SAF's humanitarian assistance after the tsunami.
15. Boey, op. cit., pp. 26–27.
16. Ibid., p. 40.
17. Interview on 19 September 2006 with Colonel Tan Chuan Jin, Assistant Chief to the General Staff (Plans) and Commander of the Humanitarian Assistance Support Group (HASG) to Meulaboh in December 2004. Interview was carried out at the Ministry of Defence (MINDEF).
18. Boey, op. cit, pp. 51–52.
19. Interview with Colonel Tan Chuan Jin, op. cit.
20. "SAF Team Sails Home after Completing Meulaboh Relief Work", *ChannelNewsAsia.com*, 22 January 2005.
21. See also "Help: Why S'pore Reached Out", *Straits Times*, 15 January 2005; and "Setting Up a Lifeline to Aceh", *Straits Times*, 20 February 2005.
22. Boey, op. cit., p. 77.
23. "SAF Mercy Flights are a Life-Saver", *Straits Times*, 12 January 2005.
24. Boey, op. cit., p. 77.
25. "Setting up a Lifeline to Aceh", *Straits Times*, 20 February 2005.
26. See SAF's Mercy Flights are a Life-saver", op. cit.; "Setting up a Lifeline to Aceh", *Straits Times*, 20 February 2005.
27. "Aid United: Strength in Numbers", *Straits Times*, 15 January 2005.
28. Information obtained from interview with Mr Lim Theam Poh, Manager, International Services; Col Christopher Chua (Retd), Senior Manager, Operations, and Ms Fara Roslan, Project Manager, TRFC, on 20 September 2006 at the Singapore Red Cross Society Building.
29. Ibid.
30. Ibid.
31. Interview with Mr Izuan M. Rais, Head, International Relief Division, and Mr Mohd Tahar Jumaat, Disaster Response (Operation), International Relief Divison, Mercy Relief, op. cit.
32. Interview with Colonel Tan Chuan Jin, op. cit.
33. Ibid.
34. Boey, op. cit., p. 64.
35. For a discussion of the rescue and relief efforts of the SAF in other areas affected by the tsunami, please see Boey, op. cit.
36. "Roundup: Singapore Plays Active Role in Tsunami Relief Efforts", *Xinhua News Agency*, 17 January 2005.
37. Interview with LTC Yazid Abdullah, Director Training Department, HQ Singapore Civil Defence Force on 14 September 2006.
38. "Mission Accomplished… and what the Operation Really Accomplished", *Straits Times*, 27 February 2005.
39. "Roundup: Singapore Plays Active Role in Tsunami Relief Efforts", *Xinhua News Agency*, 17 January 2005.
40. "PUB Puts Together Water Relief Measures to Help Tsunami-Hit Countries", News/Events, 3 January 2005, <http://www.pub.gov.sg/info_center/press_release03012005.aspx>. Information retrieved on 26 September 2006.
41. "Aid United: Strength in Numbers", *Straits Times*, 15 January 2005; "Singapore

Meshes Tsunami Relief Efforts", *Straits Times*, 28 January 2005; "Tsunami Aid Agencies and their Credentials", *Straits Times*, 15 January 2005.
42. "Tsunami Aid Agencies and their Credentials", op. cit.
43. Ibid.
44. "S'pore Firms Pitch in to Help Victims", *Business Times*, 29 December 2004.
45. "Roundup: Singapore Plays Active Role in Tsunami Relief Efforts", op. cit.
46. "ASEAN Leaders Plan Special Summit on Quake, Tsunami Disaster", *Kyodo News*, 30 December 2004; and Michael Richardson, "Tsunami Summit: Pooling Resources to Aid Everyone", *Straits Times*, 6 January 2005.
47. "After the Tsunami Summit — What Next", *Straits Times*, 8 January 2005; and "World Leaders Gather for Tsunami Summit, Aid Pledges Near US$4 Billion", *ChannelNewsAsia.com*, 6 January 2005.
48. "Singapore Pledges Another US$10m for Tsunami Efforts", *ChannelNewsAsia.com*, 6 January 2005.
49. "Indonesia Blueprint for Rebuilding Aceh", *Straits Times*, 19 January 2005.
50. Indonesia: Rebuilding a Better Aceh and Nias — Sixth Month Report, The World Bank Group. 25 June 2005. <http://www.reliefweb.int/rw/RWB.NSF/db900SID/KHII-6DU466?OpenDocument>. Retrieved on 30 September 2006.
51. BRR and International Partners, op. cit., p. 18.
52. "S'pore Tsunami Aid Close to $150m", *Straits Times*, 19 February 2005.
53. Information obtained from "Frequently Asked Questions (FAQ)", <http://www.redcross.org.sg/tsunamirelief_faq_m.htm>. Retrieved on 21 September 2006.
54. Justin Teo, "Singapore's Helping Hand in Aceh Reconstruction", *Asian Journal*, Radio Singapore International (RSI), MediaCorp Radio, 30 December 2005.
55. See "Frequently Asked Questions (FAQ)", <http://www.redcross.org.sg/tsunamirelief_faq_m.htm>. Retrieved on 21 September 2006.
56. Viability of the project in terms of a clear plan and feasible outcomes and deliverables, sustainability through the whole process of the project, exit strategy to ensure that the project outcomes can be sustained by the local community, and cooperation and synergy with local authorities/community organizations of host criteria are among some of the project evaluation criteria. For more details, please see ibid.
57. Information obtained from an interview with Mr Lim Theam Poh, Manager, International Services; Col Christopher Chua (Retd), Senior Manager, Operations, and Ms Fara Roslan, Project Manager, TRFC, op. cit., and the website of the Singapore Red Cross at <http://www.redcross.org.sg>.
58. Interview with Mr Lim Theam Poh, Manager, International Services; Col Christopher Chua (Retd), Senior Manager, Operations, and Ms Fara Roslan, Project Manager, TRFC, op. cit.
59. Ibid.
60. Information supplied by Ms Fara Roslan, Project Manager, TRFC, Singapore Red Cross.
61. "Temasek Gives US$10m Boost to Relief Efforts", *Business Times*, 7 January 2005.
62. "Temasek's Relief Efforts in Asia's Tsunami Disaster", *PR Newswire Asia*, 10 January 2005; and Temasek Holdings

Earmarks $16.5m for Relief Work", *Straits Times*, 7 January 2005.

63. "Temasek Gives US$10m Boost to Relief Efforts", op. cit.
64. This was in partnership with the Raleigh Singapore and Nias' provincial government agencies. SIF/Corporate Affairs — Sponsorships, "Indonesia Projects", <http://www.sif.org.sg/ca/ca_sponsorships_indonesia.html>. Retrieved on 27 September 2006. My appreciation to Ms Teo Hung Noi, Assistant Director, Corporate Communications of the Singapore International Foundation (SIF) for providing a write-up on "Updates of SIF Tsunami Projects in Tsunami-Affected Communities, September 2006. Source: E-mail and phone communication between author and Ms Teo on 21 September 2006.
65. "Singapore to Donate Mobility Aids to Aceh Tsunami Victims", *ChannelNewsAsia.com*, 12 August 2005; and "Updates of SIF Tsunami Projects in Tsunami-Affected Communities, September 2006. Source: E-mail communication between author and SIF on 21 September 2006, op. cit.
66. See SIF/Corporate Affairs — Sponsorships, "Indonesia Projects", <http://www.sif.org.sg/ca/ca_sponsorships_indonesia.html>. Retrieved on 27 September 2006; and "Updates of SIF Tsunami Projects in Tsunami-Affected Communities, September 2006". Source: E-mail communication between author and SIF on 21 September 2006, op. cit.
67. Ibid.
68. Donors include Panalpina World Transport (S) Pte Ltd, Cargill Asia Pacific Ltd., Neptune Orient Lines and the Tidal Waves Asia Fund with each vessel estimated to cost US$150,000. See "Singapore to Provide Fishermen in Aceh and Nias with Modern Vessels", 30 March 2005, <http://www.mercyrelief.org/fp_03.html>. Retrieved on 18 April 2006.
69. See "Tsunami-hit Aceh Medical Staff here for Training", *Straits Times*, 13 September 2005 and List of Tsunami Approved Projects (Tidal Waves Asia Fund) at <http://www.redcross.org.sg/tsunamirelief_projects_m.htm> for more details. Information retrieved on 21 September 2006 by author.
70. My deep appreciation and thanks to Mr Abdul Wahab, Liaison Officer Meulaboh, of Mercy Relief, for bringing me around Meulaboh during my field visit in August 2006. All of the interviews and discussions with officials and people on the ground were made possible through his arrangements and contacts. My thanks also go to both Mr Abdul Wahab and Mr Edward Langton, Project Manager Meulaboh, Mercy Relief, for their kind hospitality and for accommodation at Rumah Temasek during my stay there.
71. See List of Tsunami Approved Projects (Tidal Waves Asia Fund) at <http://www.redcross.org.sg/tsunamirelief_projects_m.htm>.
72. "New Meulaboh Pier Gives Residents a Logistics Lifeline", *Straits Times*, 4 April 2006.
73. "Aceh Pier Funded by S'pore Handed Over", *Business Times*, 4 April 2006.
74. "Upgrade for Hospital with Singapore Help", *Straits Times*, 4 April 2006.
75. Information obtained from interview with Mr Rudi Ariyani, Administrator of Logistics and Training, Meulaboh General Hospital (Rumah Sakit Umum Cut Nyak Dhien), Meulaboh, Aceh Barat, on 16 August 2006.

76. "Upgrade for Hospital with Singapore Help", op. cit.
77. Interview with Dr Haris M. Saputra, Director, Badan Pengelola Rumah Sakit Umum Cut Nyak Dhien Meulaboh (Meulaboh General Hospital) on 16 August 2006 in Meulaboh, Aceh Barat.
78. Interview with Mr Rudi Ariyani, Administrator of Logistics and Training, Meulaboh General Hospital (Rumah Sakit Umum Cut Nyak Dhien), op. cit.
79. "Mercy Relief to Help Rebuild Two Other Schools in Aceh Barat", 25 September 2005, <http://www.mercyrelief.org/fp_20.html>. Retrieved on 18 April 2006.
80. Interview with Mr Edward Langton, Project Manager Meulaboh, Mercy Relief, on 17 August 2006.
81. The living quarters of the Babussalalm Boarding School were quite dismal. The orphans were still living in the same building that was badly destroyed during the tsunami, their new buildings are still in the process of being built and have stalled because of a lack of funding.
82. "Mercy Relief to Help Rebuild Two Other Schools in Aceh Barat", op. cit.
83. Interview with Mr Abdul Wahab, Liaison Officer Meulaboh, Mercy Relief on 15 August 2006.
84. A *surau* at the boys orphanage has also been built and completed on 7 August 2005 (funded by the Jemaah-jemaah Masjid Al-Istighfar in Singapore).
85. Interview with Mr Yong Teck Meng, National Director, Habitat for Humanity Singapore on 13 September 2006.
86. "Over Half of $83m in Tsunami Help Given Out", *Straits Times*, 11 December 2005.
87. Interview with Ms Juliah Bee, Project Liaison Officer, Operation Hands United — Meulaboh, Habitat for Humanity Singapore Office in Meulaboh on 18 August 2006.
88. Ibid.
89. For Habitat Singapore, housing costs now stand at US$5,000 to 6,000 for a 45-square metre home.
90. Brief interview with Bapak Bupati Aceh Barat, Doctorandus Sofayanis, on 17 August 2006 at the Muhammadiyah Orphanages Presentation Ceremony for Orphans on Indonesia's National Day Celebrations.
91. Interview with Drs Hasan Abdullah, Assistant to Economic Development and Reconstruction (*Asisten Bidang Pembangunan dan Ekonomi Setdakab Aceh Barat*), on 16 August 2006, at the office of Bupati Aceh Barat.
92. Interview with Mr Tueku Ahmad Dadek, District Officer of the Administration Department, on 16 August 2006, at the office of the Bupati Aceh Barat.
93. Brief interview with Haji Zaini Bashar, Head of the Muhammadiyah, on 17 August 2006 at the Muhammadiyah Orphanages Presentation Ceremony for Orphans on Indonesia's National Day Celebrations.
94. Interview with Mr Syed Fadhlain, Muhammadiyah Youth Leader, at a coffee shop in Meulaboh on 16 August 2006.
95. Based on the author's face to face contact with the people of Meulaboh.

Concluding Remarks

Aris Ananta
Lee Poh Onn

Nanggroe Aceh Darussalam (NAD) is one of the resource-rich provinces in Indonesia. Yet, the poverty rate was as high as 29.76 per cent in 2003. It was also a region where political and armed conflicts between GAM (Gerakan Aceh Merdeka, or Free Aceh Movement) and the Government of Indonesia were commonplace for over 30 years. Because of the conflict, foreigners were not allowed to enter NAD which was therefore relatively isolated from the outside world.

Then, sequential disasters devastated the province, destroying much of the already established economic, social, and environmental surroundings. First, an earthquake struck followed by tsunami on 26 December 2004. This was a very large natural disaster, the fourth most powerful one seen in one generation. Within 45 minutes, the tsunami hit Aceh and swept over 800 kilometres of its coastal line.[1]

Three months later, on 28 March 2005, another earthquake destroyed Simeulue and Southern Aceh, as well as Nias in the neighbouring province of North Sumatra.

However, right after the disaster in 2004, an outpouring of goods, money, and labour from all over the world entered NAD to help in the relief, recovery, and reconstruction efforts of

Facing page: Housing construction in Banda Aceh. *Photo courtesy of Mercy Relief.*

the region. NAD has since been opened to international communities.

So far, much progress is visible throughout the province. Though the people in NAD still need emergency relief, the current focus is already on reconstruction, with a heavy emphasis on rebuilding housing, health facilities, and agrarian livelihood of the communities. Most children have returned to school; and health centres have reopened. Most farmers have been working on their land, and most fishing boats have been built or replaced. Indeed, compared to the world where similar natural disasters have occurred, the pace of reconstruction in NAD has been relatively and remarkably fast.

Cheerful children anticipating a brighter future in Aceh.

Photo courtesy of Mercy Relief

Furthermore, the natural disaster also resulted in the signing of an important peace accord in Helsinki on 15 August 2005, between the Government of Indonesia and the GAM. Under this accord, GAM now receives greater autonomy and the right to form political parties.[2] In return, GAM must surrender their demand for secession from the Republic of Indonesia. Therefore, the accord has provided the greatest opportunity to create a lasting peaceful recovery, reconstruction and long-term development in the province. It is almost impossible to achieve sustainable development without security and political stability. In is hoped that the sustainable development in the region will in turn help the former GAM members to integrate better into society in the province of NAD.

As a follow up to the accord, the Law on the Governing of Aceh was passed on 11 July 2006, to pave the way for the peaceful local elections (*pilkada-pemilihan kepala daerah*) conducted on 11 December 2006 to directly elect the governor and heads of districts

Photo courtesy of Lee Poh Onn

A local resident sets up a coffeeshop — back in business and ready to serve customers.

simultaneously.[3] Furthermore, one year after the peace accord, some former GAM members have been able to adjust to normal life, doing business or joining the government sector.[4] Just about two weeks before the local elections, the people seemed to be happy with the return of peace and security in NAD, but they were still frustrated with the effort to reduce poverty and correct regional imbalances within NAD.[5]

More challenges still remain. First, the huge magnitude of the disaster has left people, both the survivors and those who want to provide assistance, frustrated. Both groups would like to see

Photo courtesy of Lee Poh Onn

A tranquil village compound: renewal and peace.

things moving at a faster pace. Second, the large inflow of the money has significantly helped the region but it has also produced an inflation rate higher than the national level. During the four months after the tsunami, the CPI (consumer price index) in Banda Aceh soared by 14.7 per cent, mostly because of the destroyed transportation system and supply constraint of materials. Materials for construction and workers' wages became more expensive. The heavy cut in the fuel subsidy at the national level, carried out in October 2005, also worsened the inflation rate in this province.

At the same time, the construction boom will end as soon as the reconstruction project is over and, therefore, new earning and employment projects will be needed to replace and enhance the engine of growth generated by the construction boom. Otherwise, the growth of the economy will not be sustainable in the longer term.

Politically, the challenges include issues such as solving the difference between Jakarta and NAD on the concept of self-government, on the ideas to break the province into three provinces, on integration of GAM members, and on the use of the money poured into the province.

Nevertheless, the tsunami has changed NAD, from a relatively internationally isolated region to one open to assistance and people from all over the world. The people now have a choice on their future. They are endowed with a strategic geographical location, given its easy access to many of the most advanced economies; and rich natural assets, including gas and oil as well as beautiful beaches and parks. With the opening of the province to the world, they now also have the financial and human capital in addition to the strategic geographical location and rich natural resources. With the peace accord, the Law on Governing of Aceh and the new governor as well as the new heads of districts, they can concentrate better on long-term development and peace. The effort to rehabilitate, reconstruct and develop by and for the whole community is a golden opportunity to maintain and strengthen the momentum.

Indeed, peace has returned to NAD, strengthening the strong endowment of the province — the multi-ethnicity of the population, strategic geographical location, rich natural resources, and the current openness of the province. Therefore, the recovery, reconstruction and long-term development should cover all the population and integrate both natural disaster and conflict areas, with an attention to gender issues. If the people and the newly elected governments can manage the challenges, NAD will be facing a new dawn, with a golden opportunity for a better future to be a region of lasting peace and prosperity. As acknowledged in the BRR NAD-Nias publication,[6]

> "... now is an opportune moment to initiate wide public consultation about the direction the people prefer to take".

Notes

1. Information on the extent of the earthquake and tsunami drawn from the following reports: David Boey, *Reaching Out: Operation Flying Eagle SAF Humanitarian Assistance after the Tsunami* (Singapore: Ministry of Defence and SNP International, 2005); and BRR and International Partners, *Aceh and Nias One Year After the Tsunami: The Recovery Effort and Way Forward* (Indonesia: BRR and International Partners, December 2005).
2. Local political parties are banned in other regions in Indonesia.
3. The implementation of *pilkada* is one important step in the process of democratization in Indonesia. In 2004, Indonesia conducted the first direct elections for the post of President of the Republic of Indonesia. Since 2005, the Government of Indonesia has conducted the first series of direct elections of regional heads (governors, heads of districts, and heads of sub-districts).
4. "*Habis Bedil, Terbitlah Bisnis*" *TEMPO*, 21–27 August, 2006.
5. BE Satria "Jajak Pendapat KOMPAS. Problem Ekonomi Menghadang NAD", *KOMPAS*, 27 November 2006. The survey was conducted by KOMPAS on a sample of 753 respondents aged 17 years and over during 22–23 November 2006. A random sampling was taken from the recent telephone book directory in Banda Aceh, Sigli, Bireuen, Lhok Seumawe, Langsa, Takengon, Kutacane, Meulaboh, and Tapak Tuan. The survey does not cover those who did not have telephones, and therefore the result may reflect the opinion of the relatively better-off from society.
6. BRR NAD-NIAS, *Aceh and Nias One Year after the Tsunami. The Recovery Effort and Way Forward* (Banda Aceh, Indonesia: Badan Rehabilitasi dan Rekonstruksi NAD-Nias, 2005), p. 25.

Index

INDIA
Andaman Islands
Colombo
SRI LANKA
Nicobar Islands
MALDIVES
Banda Aceh
Meulaboh
INDIAN OCEAN
Nature's fury
unleashed